XVII Encuentro de Fisica (Physics Meeting EPN 2021)

Journal of Physics: Conference Series Volume 2238

Online
25 – 29 October 2021

ISBN: 978-1-7138-5493-7
ISSN: 1742-6588

Printed from e-media with permission by:

Curran Associates, Inc.
57 Morehouse Lane
Red Hook, NY 12571

Some format issues inherent in the e-media version may also appear in this print version.

Institute of Physics
Dirac House, Temple Back
Bristol BS1 6BE UK

Phone: 44 1 17 929 7481
Fax: 44 1 17 920 0979

techtracking@iop.org

Additional copies of this publication are available from:

Curran Associates, Inc.
57 Morehouse Lane
Red Hook, NY 12571 USA
Phone: 845-758-0400
Fax: 845-758-2633
Email: curran@proceedings.com
Web: www.proceedings.com

TABLE OF CONTENTS

ESCUELA POLITÉCNICA NACIONAL

Departamento de Física

XVII Encuentro de Física

October 2021

Journal of Physics: Conference Series **2238** (2022) 011001 doi:10.1088/1742-6596/2238/1/011001

Dear Participants,

On behalf of the organizing committee, I would like to express our gratitude for sharing and discussing the results of your research in the framework of the XVII Encuentro de Física.

October 2021 was an exciting month, in which we organized five practical workshops prior to the conference. The topics for each of these workshops were mechanics, electromagnetism, optics, biophysics, scientific programming, and the teaching of physics. All in all, more than 170 students took part in these.

During the conference, we had 13 truly outstanding keynote speakers, 52 oral presentations, and 26 poster contributions. In this edition of the XVII Encuentro de Física we decided to accept works from all the major topics of physics reflecting the interests of our Physics Department. In Fig. 1, the topics distributions of the presented works as oral contributions and posters are shown.

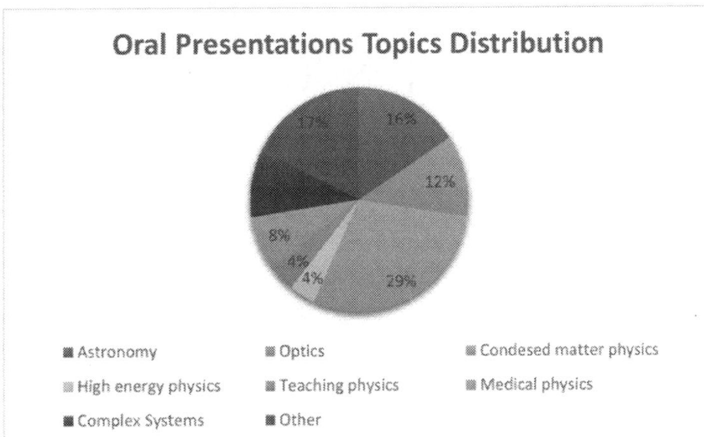

(a)

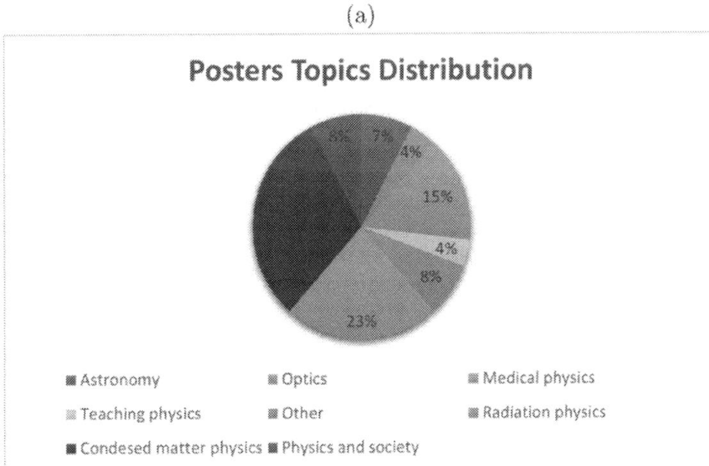

(b)

Figure 1: a) oral contributions Topics distribution b) Posters topics distribution

The authors and keynote speakers of the conference came from 15 countries of America and Europe (Fig. 2). We were pleased to witness the participation of former students of our own institution and of researchers from across the globe. I think that the Encuentro de Física can be firmly declared an international conference.

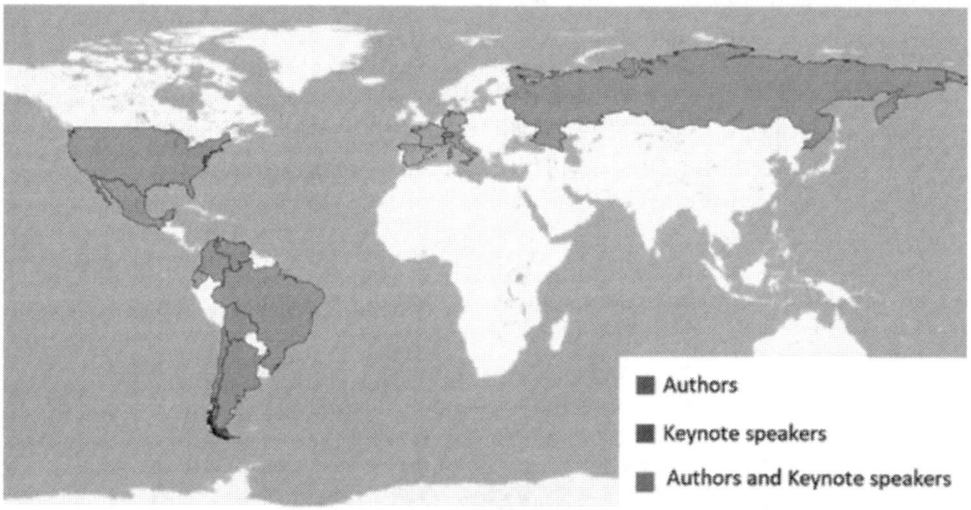

Figure 2: Countries the authors and keynote speakers come from.

Concerning the audience, 723 persons registered for the event, and more than 150 participated actively in more than 70% of the conference. We had participants from different countries of Latin America. In the following word salad figures (Fig. 3 to 5) it is possible to observe the countries, institutions, and occupations of the participants.

Figure 3: Countries/regions the participants of the XVII Ecuentro de Física come from.

Figure 4: Institutions to which the participants of the XVII Ecuentro de Física are affiliated.

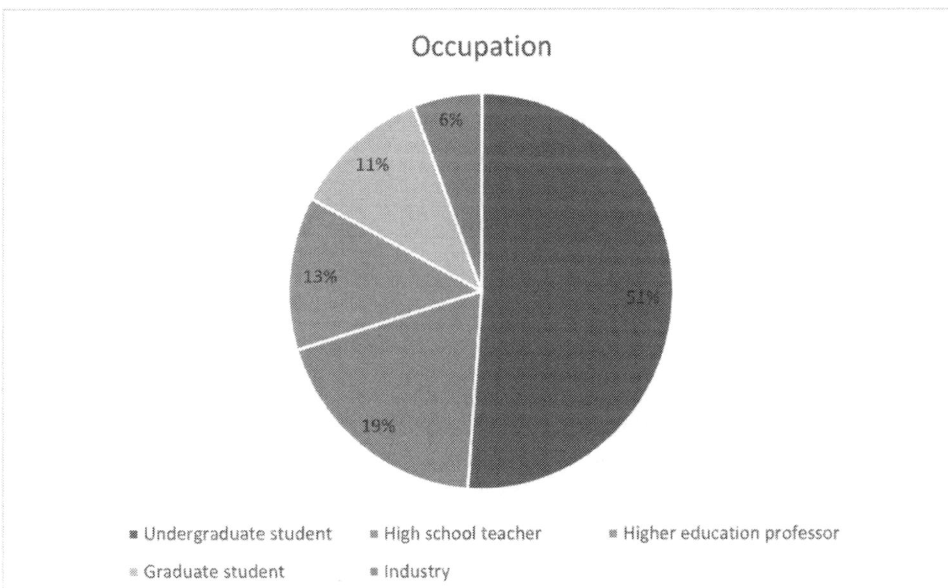

Figure 5: Professional occupations of the participants of the XVII Encuentro de Física

After the conference, we conducted a poll to evaluate the satisfaction of the participants. It is interesting to know that 67% of the people would like to keep a virtual version of the conference. Also, it should be considered for future events, that for 80% of the participants, the publication of the proceeding in an indexed by SCOPUS journal was very important.

Finally, we are pleased to mention that the overall satisfaction of the participants of the event is 4.53/5.

In the figures 6 and 7, we highlight the positive and negative aspects of the conference from the point of view of the authors and the attendees.

Figure 6: Positive aspects of the conference

Figure 7: Negative aspects of the conference

For the poster session our colleagues from the Escuela Politécnica Nacional, Dr. Edy Ayala, Dr. Luis Lascano, and Dr. Henry Osorio, evaluated the works and selected the following authors for the Best Poster Award:

- Ramazan Nazhmudinov, Observation of x-rays during heating a pyroelectric crystal by an infrared laser, Belgorod National Research University.

- Maria Isabel Subia Potosí, Quantitative analysis and simulation of a liquid crystal organizational model in human liver tissue, Escuela Politécnica Nacional.
- Hillary Estefania Rodriguez Lucas, Obtención de nanohojas de disulfuro de molibdeno y su caracterización morfológica y eléctrica, Escuela Politécnica Nacional.
- Andrés Ordóñez, DNA strand breaks directly induced by kev x-rays, Escuela Politécnica Nacional.
- Silvana Guitarra y Martín, Analysis of the reset transition of tin/hfo2/hf/tin reram devices, Universidad San Francisco de Quito.

In the present volume of the IOP´s Journal of Physics: Conference Series, we are publishing 13 works, which were presented at the conference. It should be noted that these works were suitably peer-reviewed. The participation in the conference alone did not guarantee or oblige the publication of the work in the memoirs.

We have arrived at the final stage of the XVII Encuentro de Física with this publication. We hope we will meet again in 2023 in Quito in person, notwithstanding, we will try to keep at least a virtual session to ensure the participation of those that will be unable to travel but want to contribute or attend.

Last but not least, I would like to express my gratitude to the organizing and scientific committees of the event, the peer evaluators, and the staff of Escuela Politécnica Nacional, who helped in every stage of the organization, the conference, and the publication of the proceedings.

On behalf of the Organizing Committee and the Department of Physics at EPN,

Esteban Irribarra
Chair of the XVII Encuentro de Física

Journal of Physics: Conference Series **2238** (2022) 011001 doi:10.1088/1742-6596/2238/1/011001

WELCOME TO THE XVII ENCUENTRO DE FÍSICA

Quito, 15 October 2021.

Dear participants,

On behalf of the organizing committee, I would like to express our gratitude for sharing and discussing the results of your actual research in the framework of the XVII Encuentro de Física.

We are convinced these venues to enhance and foster the possibilities to establish and strengthen working and friendship networks, are also the place where new ideas and projects begin, and finally, play a key role in motivating young researchers and students. This is particularly true in our Country, where such occasions are normally scarce. The Encuentro de Física has already established itself as a much-valued and expected tradition in this respect, and we are proud of having had the opportunity in this difficult year of maintaining its relevance and quality. In a great measure, this is due to your participation.

The pandemic caused by COVID-19 forced, unfortunately, to change the usual format of our conference into a virtual one. But on the other hand, this also provided obvious advantages. One of these advantages is that we could welcome the participation of 13 outstanding keynote speakers, who are well-known leaders in their fields without having to incur greater logistic problems. Additionally, it was possible to receive contributions from researchers working in several countries of Latin America and Europe, who could participate without the need of securing funds for traveling, which could represent a problem.

As mentioned, the program includes 13 keynote speakers, 52 oral, and 26 poster contributions. Prior to the conference, we have organized five practical workshops on mechanics, electromagnetism, optics, biophysics, scientific programming, and the teaching of physics. We are pleased to have been able to bring back such hands-on activities to the program.

This event is only possible thanks to the effort of the Professors of the Department of Physics at EPN, and thanks to the committed help of the students, assistants, and administrative personal of the Department and EPN. Of course, without the engagement of the researchers and public, this would neither have been not possible nor would have a purpose. Thank you!

We hope The XVII Encuentro de Física will meet your expectations and wish you a fruitful and interesting conference!

Esteban Irribarra
Chair of the XVII Encuentro de Física

HISTORY OF THE ENCUENTRO DE FÍSICA

El Encuentro de Física (The Physics Meeting) is a traditional event organized biennial by the Department of Physics of Escuela Politécnica Nacional (National Polytechnic School). The first edition of this meeting was held in 1989, and since then, every two years a new edition of the meeting has been developed.

The objectives of the Physics Meeting from the beginning have been:

Create a space to disseminate and debate academically the results of the most recent research carried out nationally and internationally in the different areas of study of physics. Establish networks between teachers, students and people interested in physics in Ecuador and the region. Develop workshops, active learning experiences in laboratories to improve the understanding of physics at both secondary and higher education levels. From its beginnings to the present date, 16 editions of the Physics Meeting have been organized and in this way, an open, participatory and integrative space has been built, through the development and generation of a critical and research academic community in the most relevant topics of physics.

Each edition of the event has focused on contemporary areas or research topics in physics. In the most recent editions, the meeting has focused on the Nanoscience Revolution (2009), Nuclear and Radiation Physics and the peaceful use of nuclear energy (2011), The Frontiers of Physics in Latin America (2013), Optics, in the context of the declaration of the International Year of Light by UNESCO (2015), Condensed Matter (2017) and Computational Physics (2019).

To promote the exchange of scientific knowledge for the benefit of students and the academic community in general, we have had the presentation of papers, keynote talks and lectures given by different invited researchers.

In the different editions of the event, the participation of renowned researchers has been counted. Some of the most outstanding national and international researchers who have been part of the different editions of the meeting, presenting lectures, have been: Dr. Pablo Mininni (Argentina, 2019), Prof. Sokrates Pantelides (USA, 2017), Dr. Katja Lindenberg (USA, 2015), Dr. Alexander Kubankin (Rusia, 2015), Dr. Silvia Braslavsky (Germany, 2015), Dr. Vanderlei Bagnato (Brazil, 2013), Dr. Ivan Schuller (USA, 2009), among others. There has also been the participation of undergraduate and graduate students from the country and the region, who have participated in poster presentation sessions and conferences with topics related to their degree work.

Journal of Physics: Conference Series **2238** (2022) 011001 doi:10.1088/1742-6596/2238/1/011001

COMMITTEE

Organizing Committee

- Dr. Esteban Irribarra
  Escuela Politécnica Nacional
  Chair

- Dra. Eliana Acurio
  Escuela Politécnica Nacional
  Co-Chair

- Dr. César Costa
  Escuela Politécnica Nacional
  Co-Chair

Scientific Committee

- Dr. Leonardo Basile
  Escuela Politécnica Nacional

- Dr. Gabriel Bilmes
  Universidad Nacional De La Plata

- Dr. Thomas Cadenbach
  Universidad San Francisco De Quito

- Dr. José Luis Paz
  Universidad Nacional De San Marcos

- Dr. Luca Sorriso-Valvo
  Swedish Institute Of Space Physics, Uppsala University

Collaborators

- Fis. Fernando Moncada

- Fis. Cristopher Erazo

- Fis. José Toabanda

- Ing. Janneth Peña
- Ing. Julio Erazo
- Tnlga. Karina Guerrero
- Mgtr. Paola Vega

KEYNOTES

ASTRONOMY

A JOURNEY NEAR THE SUN: FIRST RESULTS OF THE SOLAR ORBITER SATELLITE

DR. LUCA SORRISO-VALVO

Biography:

Dr. Luca Sorriso-Valvo studied his bachelor's and Ph.D. in Physics at the University of Calabria: Rende, Italy. His expertise focuses on data analysis and spatial plasma modeling and dynamics of complex systems. The fields that interest him are: turbulence in space plasma, solar physics, non-linear dynamics, space weather. He did his postdoctoral work at the University of Calabria and had a research stay at the University of California, Berkeley. Additionally, he was an associate professor at the National Polytechnic School from 2018 to 2020 and now is working at the Swedish Institute of Space Physics.

Abstract:

On February 10, 2020, a rocket that put the European Space Agency's (ESA) Solar Orbiter satellite into orbit around the sun took off from Cape Canaveral. With each turn, the probe gets closer to the star, so that by this date its perihelion is no longer so far from mercury's orbit. After more than a year and a half, the first measurements of the satellite have been developed, resulting in numerous scientific publications. These include the observation of nanoflares in the solar chromosphere, the observation of the magnetic field and high-energy particles in the magnetosphere of Venus, and a wide variety of waves and turbulent phenomena in the solar wind. In this talk, after a report on the state of the satellite, the most important results obtained will be presented, and the plan of observations expected in the coming years will be reviewed.

HIGH ENERGY PHYSICS

DARK MATTER AND STANDARD MODEL

DR. BRUCE HOENEISEN

Biography:

Bruce Hoeneisen obtained his PhD in Electrical engineering and Physics (1972) and his master's degree in Electrical Engineering (1970) at the California Institute of Technology while his undergraduate degree in civil and electrical engineering was obtained at the University of Chile (1968). He was a professor in physics at the EPN from 1972 to 1986, an independent consulting engineer in Quito since 1978, director of applied sciences at the Universidad San Francisco de Quito since 1988. He has been listed as a notable physicist in "Marquis Who's Who".

Abstract:

84% of the matter in the universe is "Dark Matter" of unknown origin. I present detailed measurements of the properties of Dark Matter by studying rotation curves of spiral galaxies, and independently, of the distributions of stellar masses of galaxies.

ELECTROWEAK INTERACTIONS: FROM DISCOVERY TO PRECISION TESTS OF THE STANDARD MODEL

DR. MAARTEN BOONEKAMP

Biography:

Maarten Boonekamp, Ph.D. es director de investigación en el Instituto de Investigación de las Leyes Fundamentales del Universo, IRFU, CEA. Obtuvo su licenciatura en Física Fundamental (1995) y su doctorado en DAPNIA / SPP y Universidad de París VII (1999). Es miembro de los experimentos DELPHI y ATLAS, especializado en Higgs y Física Electrodébil con más de 1000 publicaciones, de las cuales más de 50 contienen contribuciones personales importantes o destacadas. Además, es experto en calibración de electrones y fotones en ATLAS, con el objetivo de realizar mediciones precisas de parámetros fundamentales electrodébiles, en particular las masas del bosón W y de Higgs, y el ángulo de mezcla débil efectivo. En su investigación personal (producción de bosón de Higgs en eventos difractivos, interacciones fuertes, metodología de medición de parámetros electrodébiles) cuenta con 35 publicaciones (autor principal) y dos artículos de revisión invitados. De 2001 a 2003 fue investigadora en el CERN (Ginebra). En 2015 ganó el premio Joliot-Curie de la Sociedad Francesa de Física. Y obtuvo una beca de investigación de la ANR ("Agence Nationale de la Recherche") sobre física electrodébil en el LHC.

Abstract:

Through a brief historical introduction to the Drell-Yan process and the discovery of the electroweak gauge bosons, It will be summarized how the measurements of their production and decay properties allow to test such different things as the structure of the proton and the electroweak theory, and the guidance this provides in searches for yet undiscovered particles.

MACHINE LEARNING / COMPLEX SYSTEMS

ARTIFICIAL INTELLIGENCE FROM PHYSICS

DR. KONRAD KORDING

Biography:

Konrad Paul Körding, Ph.D. is a professor at the University of Pennsylvania. He obtained his undergraduate degree (1997) and PhD (2001) in physics at the Federal Polytechnic School in Zurich. He has completed post-doctoral stays at the Swiss College of Zurich and University College of London where he obtained a Position of Heisenberg Fellow at MIT. In 2017 he joined the Department of Neuroscience and also the Department of Bioengineering at the University of Pennsylvania. Currently his research work is focused on methods of neural data analysis and methods of obtaining large sets of neural data.

Abstract:

Machine learning is starting to be a workhorse of science, engineering, and society. We use it for everything from cosmology, via self-driving cars, to deciding who gets a credit card. And yet, we are still limited at understanding why the various machine learning approaches even work. In my talk I will review how machine learning is relevant across society. I will also discuss various ways from physics and mathematics for understanding how learning works. Lastly, I will argue for more integration between all these field.

COMPLEX NETWORKS AND THEIR APPLICATIONS

DR. VITALY BELIK

Biography:

Dr. Vital Belik is an assistant professor at the Institute for Veterinary Epidemiology and Biostatistics at Free University of Berlin, and he is leader of Working Group System Modelling. He obtained his diploma (with honors) in Physics and Biochemical Physics (2004) from Lomonosov Moscow State University, PhD in Theoretical Physics from Georg-August-Universität Göttingen (2008) and postdoctoral experience at Massachusetts Institute of Technology (2012). He has worked as consultant intern at D-Fine, researcher at Max Planck Institute for Dynamics and Selforganization, visiting scientist at Helmholtz center for Infection Research and as research associate at Institute for Theoretical Physics, Technische Universität Berlin. His research focuses on Microbiome dynamics, Computer vision and Machine learning in animal health applications, Disease spread on networks of hospitals and animal trade networks, Time series analysis in application to sensor data on animal behavior for health monitoring, Parasite-host-environment dynamics and Hormon dymanics in dogs. Also, he is Co-Organizer of the Berlin Colloquium Statistical Methods in Empiric Research, Representative of FU Berlin in Data Literacy Education Network of

Stifterverband and Review Editor on the Editorial Board of Frontiers in Veterinary Science COST Action Workshop organizer (Networks in One Health Epidemiology)

Abstract:

Complex networks is a vivid field of research having its roots in mathematical graph theory originating at least in the 18th century. Nowadays networks represent a suitable framework predestined for description of complex systems in a large range of spatial and temporal scales – from molecular (e.g. protein-protein interaction networks) to societal (e.g. contact networks) and planetary scale (e.g. transportation networks). Complex system are consisting of many individual units (nodes or vertices) interacting with each other. Interactions are described as edges or links between vertices. In this talk I will provide an introduction into network science and its applications especially in the field of epidemiology.

Journal of Physics: Conference Series **2238** (2022) 011001 doi:10.1088/1742-6596/2238/1/011001

MEDICAL PHYSICS

ARTIFICIAL INTELLIGENCE IN MULTIMODALITY MEDICAL IMAGING: PAST, PRESENT, ANY FUTURE?

DR. HABIB ZAIDI

Biography:

Professor Habib Zaidi is head of the PET Instrumentation & Neuroimaging Laboratory at Geneva University Hospital and faculty member at the medical school of Geneva University. He is also a Professor of Medical Physics at the University of Groningen (Netherlands), Adjunct Professor of Medical Physics and Molecular Imaging at the University of Southern Denmark, Adjunct Professor of Medical Physics at Shahid Beheshti University and visiting Professor at Tehran University of Medical Sciences. He is actively involved in developing imaging solutions for cutting-edge interdisciplinary biomedical research and clinical diagnosis in addition to lecturing undergraduate and postgraduate courses on medical physics and medical imaging. Prof. Zaidi has been an invited speaker of over 160 keynote lectures and talks at an International level, has authored over 700+ publications including 323 peer-reviewed journal articles in prominent journals, 340 conference proceedings and 41 book chapters. His academic accomplishments in the area of quantitative PET imaging have been well recognized by his peers and by the medical imaging community at large since he is a recipient of many awards and distinctions.

Abstract:

This talk presents the fundamental principles and major applications of artificial intelligence (AI), in particular deep learning approaches, in multimodality medical imaging. To this end, the applications of deep learning in five generic fields of multimodality medical imaging, including imaging instrumentation design, image denoising (low-dose imaging), image reconstruction, quantification and segmentation, radiation dosimetry and computer-aided diagnosis and outcome prediction are discussed. Deep learning algorithms have been widely utilized in various medical image analysis problems owing to the promising results achieved in image reconstruction, segmentation, regression, denoising (low-dose scanning) and radiomics analysis. This talk reflects the tremendous increase in interest in quantitative molecular imaging using deep learning techniques in the past decade to improve image quality and to obtain quantitatively accurate data from dedicated combined PET/CT and PET/MR systems. The deployment of AI-based methods when exposed to a different test dataset requires ensuring that the developed model has sufficient generalizability. This is an important part of quality control measures prior to implementation in the clinic. Novel deep learning techniques are revolutionizing clinical practice and are now offering unique capabilities to the clinical medical imaging community. Future opportunities, the challenges facing the adoption of deep learning approaches and their role in molecular imaging research are also addressed.

Journal of Physics: Conference Series **2238** (2022) 011001 doi:10.1088/1742-6596/2238/1/011001

GEL DOSIMETRY, APPLICATIONS OF MACHINE LEARNING IN RADIOTHERAPY

DRA. JULIANA PAVONI

Biography:

Juliana Pavoni is an assistant professor at the Department of Physics, Faculty of Philosophy, Science and Letters of Ribeirão Preto - University of São Paulo and works as a clinical medical physicist in radiotherapy at the Ribeirão Preto Medical School Hospital and Clinics - University of São Paulo. She received a BS degree in medical physics (2004) and a Ph.D. degree in Physics Applied to Medicine and Biology (2009) from the University of São Paulo. She has experience in Medical Physics, with a particular interest in radiopherapy quiality control tools, including gel dosimetry and artificial intelligence.

Abstract:

The use of machine learning and other sophisticated models to aid in prediction and decision-making has become widely popular across the medical field. The situation is not different in radiotherapy. These resources allow automation and workflow optimizations, especially in automatic segmentation, planning, and quality assurance. In this presentation, an overview of the application of these tools in radiotherapy will be discussed.

Journal of Physics: Conference Series **2238** (2022) 011001 doi:10.1088/1742-6596/2238/1/011001

OPTICS

LOW-COST METHODS AND INSTRUMENTS FOR ADVANCED HOLOGRAPHIC MICROSCOPY

DR. AYDOGAN OZCAN

Biography:

Dr. Aydogan Ozcan received his Ph.D. degree from the Department of Electrical Engineering at Stanford University. He joined UCLA in 2007 and is currently Chancellor Professor and HHMI Professor at the Howard Hughes Medical Institute. He holds more than 45 registered patents (and more than 20 in process) for his inventions in telemedicine, mobile health, nonlinear optics, optical fibers, optical coherence tomography, etc. He has given more than 70 plenary lectures and more than 350 invited talks. He is the author of a book and co-author of more than 700 peer-reviewed publications in major journals and conferences. He has received numerous awards, most notably in 2011 when he received the Presidential Young Career Award for Scientists and Engineers (PECASE), which is the highest honor awarded by the U.S. government in science and engineering professionals in early stages of their research careers.

Abstract:

We will discuss diffractive optical networks that can all-optically implement various complex functions, as the input light diffracts through spatially-engineered surfaces. These diffractive processors will find applications in all-optical image analysis, feature detection and object classification, also enabling task-specific camera designs and new optical components.

TEACHING OF PHYSICS

TEACHING OF PHYSICS

DR. ERIC BURKHOLDER

Biography:

Eric Burkholder is an assistant professor of physics at Auburn University in the United States. He runs a research group studying a wide range of phenomena in physics and engineering education including real-world problem solving, student perceptions of physics tasks, student motivation, social psychological factors, and group work. He previously completed a postdoctoral stay in Physical Education at Stanford University under the direction of Nobel Laureate Carl Wieman. He holds a Ph.D. in Soft Matter Physics from the California Institute of Technology and an undergraduate degree in Chemical Engineering from Cornell University.

Abstract:

Traditional university education has remained largely unchanged since medieval times – a professor standing in front of their students telling the students the information they are supposed to learn. Over the past few decades, research has consistently demonstrated that this is a poor method of teaching because it does not actively engage students in the learning process. "Active learning" – any method of teaching in which students are engaged by solving problems, talking to each other, answering conceptual questions, etc. instead of listening to the teacher – has been shown to help students learn new content, decrease failure rates, and improve student's perceptions of science. There are many different ways of making a classroom more active, but all methods are based on fundamental principles of learning that have been investigated by cognitive scientists. In this talk, I will give a brief overview of the most important principles of learning and discuss practical ways to implement this in the physics classroom. As we go, I will present data showing how these novel methods lead to improved student outcomes. At the end of the talk, I will discuss what factors are necessary to facilitate systemic change in physics teaching. Hopefully, this will inspire individual faculty to adopt more effective teaching practices and encourage administrators to rethink the way teaching is evaluated and valued at the university.

TEACHING OF PHYSICS

DRA. DIANA LÓPEZ TAVARES

Biography:

Diana López obtained her PhD in Physics Education Research from the National Polytechnic Institute of Mexico. She has a master's degree in Physics Education Research and an undergraduate degree in Physics. He has collaborated with PhET since 2017 working on the design of physics simulations and teaching professional design to Hispanic professors. Her research focuses on advanced learning in Science, Technology, Engineering and Mathematics (STEM) for students in Latin America and the world using equitable education of the highest quality, supporting programs for the professional development of teachers and designing contextualized study material. She currently works as a Latin American specialist in the PhET project and during the pandemic has focused on the design and implementation of massive virtual courses and virtual teaching activities to enhance remote classes of science teachers.

Abstract:

The PhET project from the University of Colorado Boulder creates interactive simulations and free virtual labs for science and math learning. PhET simulations are used globally, and due to the COVID-19 pandemic, in some regions of the world their use increased by up to 700%, being one of the main resources used by science teachers. It is almost 20 years of the PhET Project and in this conference we will review some of the research results on the impact on student learning and motivation when using simulations in class, the research that is still open in PhET, and the projects that are held in Latin America with the PhET-Global initiative.

LOW-COST EXPERIMENTAL RESOURCES IN THE TEACHING OF PHYSICS TO SHARE WITH TEACHERS

DR. JOSÉ BENITO VÁZQUEZ DORRÍO

Biography:

José Vásquez obtained a degree in Physical Sciences from the University of Santiago de Compostela in 1990 and a PhD in Physics from the University of Vigo in 1996. He has been awarded four six-year periods of CNEAI research and his research in the Engineering Physics Group focuses on applied optics, photonics, and laser technologies. He is co-author of more than 70 publications indexed in SCOPUS, plus 50 publications in congresses and has collaborated in the organization of 14 scientific meetings. He maintains a line of educational innovation in the field of experimental science teaching since 1991 with 10 innovation and scientific transfer projects, 120 publications and 38 participations in the organization of scientific meetings. He currently directs a doctoral thesis, is a regular censor of 5 international journals and is vice president of the Hands-on Science Association. In addition, he is Ambassador for Spain of Scientix and head of the Teaching Innovation Group of Manipulative Learning of Physics.

Abstract:

Almost all reports on education, learning, dissemination, and vocation in the teaching of physics indicate that it is necessary to establish links between the contents of formal learning and the experiential learning inherent in the scientific method. An obstacle that appears when it is intended to carry out this necessary connection lies in the tendency of teachers to replicate in the classroom the teaching that they have received during their student stage. This teaching often focuses on the theoretical foundations and moves away, in most situations, from an active, practical, and experimental vision. The ability to teach using manipulative, practical, experimental, and material-focused activities should be an objective of the training of teachers who should be offered the possibility of developing competences in this field.

This talk presents an updated review of the possible resources that, in this constructivist context, teachers can use. Different possibilities of use outside and inside the classroom are analyzed, as well as recent results of various training courses carried out for teachers with this methodology. The purpose of the course is to provide participants with an overview of the advantages and possibilities of using experimental activities in formal education, inducing, if possible, their use in their daily teaching work, so that students can "do" science rather than simply being "exposed" to it.

Journal of Physics: Conference Series	**2238** (2022) 011001	doi:10.1088/1742-6596/2238/1/011001

SOLID STATE

PEROVSKITE SOLAR CELLS

DR. WOLFGANG TRESS

Biography:

Dr. Wolfgang Tress is currently a full professor at the University of Applied Sciences in Zurich. He studied at the Technical University of Dresden. He is the author of the renowned book "Organic Solar Cells: Theory, Experiments and Simulation of Devices" (2014). He has received multiple awards among which it is highlighted that, in 2016, he was awarded the Zeno Karl Schindler prize in the Field of Environmental Sciences and Sustainability thanks to his postdoctoral work. Wolfgang was a Marie-Curie fellow at the Ludwig Maximilian University in Munich and a Ambizione Fellow at the École polytechnique fédérale de Lausanne. He is also a member of the editorial board of the open access journal "Research in Advanced Energy and Sustainability". His research is focused on the physics of perovskite solar cell devices; and more recently he has investigated phenomena of recombination and hysteresis in this system.

Abstract:

In the last decade, metal-halide perovskite semiconductors have received tremendous attention in research due to their excellent optoelectronic properties, making them highly interesting materials for solar cells and light emitting diodes (LEDs). It is fascinating that these perovskites are highly tolerant against electronic defects, which allowed for solar-cell efficiencies greater than 25% and at the same time show ionic conductivity meditated through mobile lattice defects.

In this talk the device physics of perovskite solar cells is discussed. Focus is on the peculiar properties of the material, which enable low recombination losses and thus high photovoltages. Furthermore, the various effects of ion migration on device performance are described. They range from hysteresis in the current-voltage curve of solar cells, to reversible degradation during long-term operation, to high gains in photodetectors and the electroluminescence in LEDs that depends on the mode of operation.

Key in understanding these phenomena is the interplay between ionic and electronic conductivity, where the ionic response belatedly modifies the electronic response, which is the one commonly observed in devices.

This interplay has consequences on how to interpret the results of common characterization techniques. In addition, the real-world long-term operation of perovskite optoelectronic devices under varied ambient conditions is affected. These points lead towards the open challenges, which conclude the talk.

Journal of Physics: Conference Series **2238** (2022) 011001 doi:10.1088/1742-6596/2238/1/011001

A BRIEF LOOK INTO ELECTRON MICROSCOPY: UNVEILING MATERIALS PROPERTIES AT THE ATOMIC AND NANOMETER LEVEL

DR. JUAN CARLOS IDROBO

Biography:

Juan Carlos Idrobo is a Senior Staff Scientist at Oak Ridge National Laboratory and Group Leader of the Scanning Transmission Electron Microscopy Group at the Center for Nanophase Materials Sciences. His research consists in applying analytical techniques in electron spectroscopy within monochromated and aberration-corrected scanning transmission electron microscopy to study the structure and electronic, magnetic, thermal, optical and topological properties of materials. He has published over 170 papers in peer review journals with over 16,000 citations. In 2018 Idrobo was recognized by Clarivate Analytics to be among the top 1% of researchers being cited in Cross-Field in Web of Science, between 2006 and 2016. Idrobo holds Physics degrees from Universidad de Los Andes in Colombia (B.Sc., 2000), University of Illinois at Chicago (Master, 2003) and University of California Davis (Ph.D., 2004).

Abstract:

Modern scanning and transmission electron microscopes (S/TEM) are now almost ubiquitous in materials and biological sciences laboratories. They have radically enhanced our understanding of organic and inorganic matter with the successful development of aberration correctors [1,2], detectors with film-equivalent dynamical range [3], and more recently, with monochromators capable of achieving sub-10 meV energy resolution spectroscopy [4].

Here, I will present several examples demonstrating how we have exploited these capabilities and solved the pertinent experimental challenges to probe materials behavior at the nanometer and atomic scales. Specifically, I will show how by utilizing the phase of the electron probe one can reveal the magnetic order of complex-oxide materials at the atomic level. I will also explain how the new generation of monochromators, combined with aberration-corrected STEM, can be used (i) as a primary thermometer (without requiring any previous knowledge of the sample) [5].

I will also explain how the new generation of monochromators, combined with aberration-corrected STEM, can be used (i) as a primary thermometer (without requiring any previous knowledge of the sample) [6]; and to detect site-specific isotopic labels in amino acids at the nanometer scale [7]. Additionally, I will show how one can now visualize the electric field of individual atomic columns of heavy and light elements, detect charge density inhomogeneities and the magnetic order of materials at the sub-Angstrom scale by using an ultra-low noise SCMOS detector in the diffraction plane [8].

Lastly, I will discuss potentially relevant new challenges in electron microscopy. For instance, will it be possible to map orbitals and spins with atomic resolution and with single atom sensitivity; detect a superconducting transition; measure the specific heat and thermal conductivity of materials; map Hall effects or topological properties with nanometer resolution [9]? These questions will be addressed and further elaborated during the presentation [10].

References:

1. J. Zach and M. Haider, Optik 99 (1995), p. 112.
2. O. L. Krivanek, et al, Institute of Physics Conference Series 153 (1997), p. 35.
3. A. R. Faruqi, R. Henderson, Curr. Opin. Struc. Biol. 17 (2007), p. 549.
4. O. L. Krivanek, et al, Phil. Trans. R. Soc. A 367 (2009), p. 3683.
5. J. C. Idrobo, et al, Adv. Struc. Chem. Img. 2 (2016), p. 5; J. C. Idrobo, et al, unpublished (2021).
6. J. C. Idrobo, et al, Phys. Rev. Lett. 120 (2016), p. 095901.
7. J. A. Hachtel, et al, Science 363 (2019), p. 525.
8. J. A. Hachtel, et al, Adv. Struc. Chem. Img. 4 (2018), p. 10; Q. Zheng et al, Science Adv. 7, eabe6819 (2021); F. Křížek et al, arXiv:2012.00894 (2020).
9. J.C. Idrobo, Nature Rev. Mater. 2, (2021) p. 100.
10. This research was supported by the Center for Nanophase Materials Sciences, which is a Department of Energy Office of Science User Facility, and instrumentation within ORNL's Materials Characterization Core provided by UT-Battelle, LLC under Contract No. DE-AC05-00OR22725 with the U.S. Department of Energy.

CULTURAL ACTIVITIES

A VIRTUAL TOUR OF THE GALAPAGOS ISLANDS

MSC. DANIELA COX

Biography:

Daniela Cox is a third generation from Galapagos linked to cultural and environmental initiatives of her island community from an early age. Master in Sustainable Business from Griffith University, Community Organizer in Wilderness Society Australia, Naturalist Guide of the Galapagos National Park, Municipal Environmental Director, youngest woman elected politician in Galapagos. Since 2019 he has lived in Denmark with his family, he is a member of the Nordic B Leaders Network, promoter of Regeneration, Ex Advisory Board of Women 4 Solutions, and CEO of Plura Consulting, the Galapagos-Scandinavian firm that helps individuals and organizations to integrate and experiment sustainability and regeneration from a holistic and systemic vision. Due to her multisectoral trajectory and her activism in Sustainability and Diversity, Daniela is a local and international speaker and presenter

Abstract:

Almost a thousand kilometers away from South America, Galapagos is a dynamic, unique and fascinating archipelago of the Ecuadorian territory. The natural wealth of its geography, geology and ecology harmonizes with the consciousness, resilience, and adaptability of its island population. What are the challenges and opportunities for regenerative and mindful post-pandemic tourism? What ecosystems does the marine reserve sustain for conservation and for economies in the region? And how its island history and culture are the key to building new community narratives anywhere. Join Daniela Cox in a photographic presentation with a local and global perspective on Galapagos, in an invitation to learn more about the Enchanted Islands, which inspired Charles Darwin.

ORAL CONTRIBUTIONS

Astronomy

Astronomy for development: Astrostays in the middle of the world

Nicolás Vásquez[1], Marcela J. Morillo Acosta[2], Andrés Baquero, Raúl Puebla
[1]Escuela Politécnica Nacional;[2]Universidad Técnica de Cotopaxi

---

ZTF constraints on variability from intermediate-mass black hole candidates

Ernesto Camacho[1], Paula Sánchez-Sáez, Franz Bauer, Paula Sánchez-Sáez
[1]Pontificial Catholic University of Chile

Characterising cosmic filaments using hydro-dynamical simulations

Daniela Galárraga-Espinosa[1]
[1]Institut d'Astrophysique Spatiale

Extragalactic Fast X-ray Transient Candidates Discovered by Chandra

Jonathan Quirola[1], Franz Bauer
[1]Pontificia Universidad Católica de Chile

An overview of shock-multicloud interactions in magnetised galactic winds

Wladimir Eduardo Banda-Barragán[1]
[1]Universität Hamburg

Equatorial total electron content

Ericson López[1], Erik Toapanta, Hugo Barbier[1]
[1]Escuela Politécnica Nacional de Ecuador

Evaluation of the survival probability of exoplanets orbiting low-mass stars with enhanced energy loss

Santiago Arceo Díaz[1], Elena Elsa Bricio Barrios, Kai Zuber
[1]Tecnológico Nacional de México

Complex systems

Primary cycles in complex networks

Ramon Xulvi-Brunet[1]
[1]Escuela Politécnica Nacional

Statistical Analysis Of Pseudo-Local Properties In Turbulent Space Plasmas Using 3d Numerical Simulations

Rocio Elizabeth Manobanda Guamán[1], Luca Sorriso-Valvo[2], Renaud Ferrand, Christian Leonardo Vásconez Vega
[1]Universidad Técnica de Ambato;[2]National Research Council

---

Newton-Prygogine y la inter-ciencia

David Guzmán[1]
[1]PUCE

Statistical properties of Ecuadorian seismicity

Ramon Xulvi-Brunet[1], Stephen Hernández, Luca Sorriso-Valvo, Mario Calixto Romero Ruiz[1]
[1]Escuela Politécnica Nacional

3D on-lattice stochastic reaction-diffusion simulation of protein-membrane reversible association

Lupe Villegas[1], Jose Requejo-Isidro
[1]Centro Nacional de Biotecnología (CNB), CSIC.

Condensed matter physics

Influence of dopamine on the magnetic properties of superparamagnetic iron oxide nanoparticles

Jilder Dandy Peña Serna[1], Miguel Novak, Claudio Sangregorio, Benjamin Rache Salles, Martin Albino, Oana Pascu, Davide Peddis,Oana Pascu
[1]Centro Brasileiro de Pesquisas Físicas

Electric multipole moments in crystalline insulators

Rodrigo Sandoval[1], Leonardo Basile[1], Wladimir Benalcazar
[1]Escuela Politécnica Nacional

Structural and Thermal Properties of InSb:Mn with MnSb Clusters

Jesús González-Laprea[1], Eduardo Quintana, Jaime A. Castro, José R. Fermin[2], Carlos Durante[2], Martín Emilio Mendoza Oliveros
[1]Pontificia Universidad Católica del Ecuador;[2]University of Zulia

---

Improving the electrical properties of graphene inks by controlling nanosheet thickness distribution

Jimmy Narváez[1]
[1]Escuela Politécnica Nacional

Molecular dynamics algorithm for simulating KeV particles bombardment

Ramon Xulvi-Brunet[1]
[1]Escuela Politécnica Nacional

Electrical characterization of molecular junctions using graphene obtained by electrochemical exfoliation

Esteban Humberto Yépez Jiménez[1], Henry Marcelo Osorio Calvopiña[1]
[1]National Polytechnic School

Parametric X-ray Radiation from nano meter powders

Esteban Irribarra[1], A.S. Kubankin, A. N.Eliseyev, R.M. Nazhmudinov, A.S. Kluev, V. I. Alexeyev, R.M. Nazhmudinov, Kishin Ivan[2], A.S. Kluev
[1]Escuela Politécnica Nacional;[2]Belgorod National Research University

---

ZnO thin films grown at low temperature by PE-ALD for application in electronic devices

Jhonathan Rafael Castillo Saenz[1], Marcelo Martínez Puente, Eduardo Martínez Guerra, Mario Alberto Curiel Álvarez, Francisco Servando Aguirre Tostado, Eduardo Martínez Guerra, Nicola Nedev, David Mateos, Benjamín Valdez-Salas, Oscar Perez-Landeros, María Isabel Mendivil Palma[2]
[1]Universidad Autónoma de Baja California;[2]Centro de investigación en materiales avanzados S. C.

Adsorption enhanced photocatalytic degradation of Rhodamine B using GdxBi1-xFeO3@SBA-15 (x= 0, 0.05, 0.10, 0.15) nanocomposites under visible light irradiation

Maria Jose Benitez[1], Sofia Andrade, Karla Sofía Vizuete Armendáriz[2], Valeria Ochoa, Thomas Cadenbach[3], Alexis Debut[2]
[1]Escuela Politécnica Nacional;[2]Universidad de las Fuerzas Armadas ESPE;[3]USFQ

Catalyst dependency on the growth of single-walled carbon nanotubes

Carlos Reinoso[1], Luis Corredor[1], Roberto Andrade
[1]University YachayTech

Synthesis and characterization of N-doped graphene quantum dots-titanium dioxide nanocomposite for photodegradation

Isamar Sarabia Ayala[1]
[1]National Polytechnic School

Green Synthesis to produce copper nanoparticles and functionalize with graphene oxide to evaluate antimicrobial activity.

Joseth Alejandra Pachacama Ruiz[1], Fabian Ernesto Arias Arias
[1]Escuela Superior Politécnica De Chimborazo

Modifications in the structure of graphene films deposited on glass by a 100 eV to 10 keV electron beam

Karla Estefanía Moyano Quishpe[1], Cristian Patricio Santacruz Teran[1], Cesar Costa Vera[1], Carlos Reinoso[2], Beatriz Pérez, Esteban Irribarra[1], Steven Nuñez[1]
[1]Escuela Politécnica Nacional;[2]University YachayTech

Synthesis of silicon oxide nanoparticles using an environmentally friendly approach and their application for the removal of emerging contaminants in aqueous media

Victor H. Guerrero[1], Jennifer Lorena Tejedor Oyos[1]
[1]Escuela Politécnica Nacional

Analysis of the physical mechanism behind the reset transition of HfO2-based ReRAM devices

Silvana Guitarra[1], Martín Santiago Gavilánez Aguilar[1]
[1]Universidad San Francisco de Quito

Photocatalytic activity of compounds obtained from naturally occurring minerals by an environmentally friendly process using pressurized aqueous solutions

Victor H. Guerrero[1], Patricia I. Pontón, Carla Valdivieso[2]
[1]Escuela Politécnica Nacional

IOP Publishing
2238 (2022) 011001 doi:10.1088/1742-6596/2238/1/011001

High-energy physics

Search for invisible Higgs boson decays with vector boson fusion signatures with the ATLAS detector using an integrated luminosity of 139 fb-1

Pablo Rivadeneira Bracho[1], O Munir[2]
[1]DESY;[2]Lehigh Univ

Precision bounds on composite Higgs models at the FCC-ee

Andres Pinto[1], Giacomo Cacciapaglia, Aldo Deandrea
[1]Université Paris-Saclay

Journal of Physics: Conference Series **2238** (2022) 011001 doi:10.1088/1742-6596/2238/1/011001

Medical physics

Effects of movement artifacts in nuclear hybrid modalities for image diagnostic

Pedro Fernando Escudero Villa[1]
[1]Universidad Tecnológica Indoamerica

Cross calibration of ionization chambers using linear accelerators as a standard to replace isotopic sources and their use for radiotherapy dosimetry

Josue Vallejo[1]
[1]Solca Manabí

Shielding assessment in a computed tomography facility

Washington Javier Carrasco Tuston[1]
[1]Hospital de Especialidades Jose Carrasco Arteaga

Comparisonbetween 3D-CRT, IMRT and VMAT techniquesfor prostate cancer radiation treatment

W V Ona[1]
[1]RadiotherapyUnit, Hospital de Especialidades Carlos Andrade Marin, Quito Av. 18 de Septiembre y Av. Universitaria, Ecuador

---

Optics

Square metallic shields in photonic crystals

Danny Manuel Calvo Velasco[1]
Corporación Universitaria Comfacauca

Twisted light in vacuum described as a phasor

Erick Lamilla Rubio[1], Manuel Alvarez Alvarado, Peter Iza[1], Arturo Pazmiño Velez
[1]Escuela Superior Politecnica del Litoral

Implementation And Testing Of An Optical Tweezers For Capture And Manipulation Of Microparticles

Eugenia Elizabeth Samaniego Onofre[1], Cesar Costa Vera[1]
[1]Escuela Politécnica Nacional

Modelling ultrafast laser ablation of aluminum films

Jose Camilo Diaz B[1], Dmitry Ivanov, Gabriel M. Bilmes[2], Demian Biasetti
[1]Universidad Nacional De La Plata;[2]Centro de Investigaciones Opticas (CONICET-CIC-UNLP)

Luminescent nanothermometry for biological applications

Luis José Borrero González[1]
[1]Pontificia Universidad Católica del Ecuador

TOF MS Spectra simulation for ions from ultrafast and high-field-induced molecular dissociation on the surface of nanoparticles

Cesar Costa Vera[1]
[1]Escuela Politecnica Nacional

Photonics and the development of immersive virtual environments

Daniel Alejandro Loaiza Carvajal[1], María Mercedes Morita[2], Gabriel M. Bilmes[2]
[1]Centro de investigaciones opticas -CIOp;[2]Centro de Investigaciones Ópticas (CONICET-CIC-UNLP)

Design And Construction Of A 3D Printed Portable Libs System

Omar Oporto Bernal[1], Omar Ormachea, Alex Villazón
[1]Universidad Privada Boliviana

Journal of Physics: Conference Series **2238** (2022) 011001 doi:10.1088/1742-6596/2238/1/011001

Teaching of physics

Kolb's Experiential Learning Cycle In The Teaching Of Physics, Particular Thematic: The Archimedes' Principle

Byron Giovanny Méndez Puenayán[1]
[1]Universidad Central del Ecuador

Coaque, Archeological and Astronomical Museum

Jennifer Alexandra Chacón Chávez[1], Ricardo Andrés Caiza Grijalva, Sandra Teresa Procel Guerra, Cristopher Ricardo Erazo Vallejos, Diego S Dominguez[1], Nicolás Vásquez[1], Ana Cabero del Río, Reyes Benítez Pedro Santiago
[1]Escuela Politécnica Nacional

Other

ON-state reliability of GaN-on-Si Schottky Barrier Diodes: Si3N4 vs. Al2O3/SiO2 GET dielectric

Eliana Maribel Acurio Méndez[1]
[1]Escuela Politécnica Nacional

Study of Hydrogen Sorption/ Desorption Effect on Austenitic Iron-Based Alloys

Katherine Encalada[1], Amarante Böttger
[1]Escuela Politécnica Nacional

Dust adhesion physics on photovoltaic surfaces and its application on assisted natural cleaning of solar trackers

Giovanni Alejandro Cruz Ortiz[1], Pavel Vorobiev
[1]Centro de Investigación en Materiales Avanzados

Radon Indoor Air Pollution

Jheny Orbe Ordoñez[1], Marcella Capua, Gabriela Belén Ureña
[1]Escuela Superior Politécnica del Chimborazo

Temporal Causality in Quantum Teleportation

Jiri Svozilik[1], Jonnathan Pineda Pineda[2]
[1]Palacký University, Olomouc;[2]Yachay Tech University

Design and Construction of a Prototype Machine for Thermoplastic Materials Assembly by Laser Transmission Welding

Juan Diego Jijón Valdivieso[1], Cesar Costa Vera[2]
[1]Instituto de Investigación Geológico y Energético;[2]Escuela Politecnica Nacional

---

Polarization And Biomineralization Of Hydroxyapatite-Barium Titanate Composites

Rafael Uribe[1], Gema Gonzalez, Isamara Rojas, Luis Lascano, Gema Gonzalez
[1]Escuela Politecnica Nacional

Synthesis of graphene hydrogel and its application in capacitive deionization

Ronny Omar De La Bastida Chiza[1], Leonardo Basile[1]
[1]National Polytechnic School

Influence of windy conditions in precipitation particle size distribution corrections in Antisana Region

Luis Felipe Gualco Centeno[1], Luis Maisincho, Lenin Campozano[1], Thomas Condom, Marcos Villacís
[1]Escuela Politécnica Nacional;

Journal of Physics: Conference Series 2238 (2022) 011001 doi:10.1088/1742-6596/2238/1/011001

POSTERS

Astronomy

On the low star formation rate in Sagittarius C

Karen Marina Martínez Acosta[1]
[1]Pontificia Universidad Católica de Valparaíso - Universidad técnica Federico Santa María

---

Simulation of exoplanet spectra to be measured by JWST-LRS-Slitless

BARBIER[1]
[1]National Polytechnic School

On the destruction of shocked clouds in the interstellar medium

Kevin Josué Imacaña Pozo[1], Wladimir Eduardo Banda-Barragán
[1]National Polytechnic School

Journal of Physics: Conference Series **2238** (2022) 011001 doi:10.1088/1742-6596/2238/1/011001

Complex systems

Numerical model of acoustic dispersion around oil wells in Yasuní National Park.

Geoconda Marisela Velasco Castelo[1], Christian Leonardo Vásconez Vega
[1]UTA

Condensed matter physics

Study Of Modification Of The Inner Surface Of The Hollow Poly-Sulphone Fiber Channels When Irradiated By A Beam Of Electrons With An Energy Of 10 Kev

Larisa Myshelovka[1], Kristina Vokhmyanina, Artem Pyatigor, Valentina Sotnikova
[1]Belgorod National Research University

Defect Passivation of Hybrid Perovskite Methyl Ammonium Lead Iodide with organic molecules: The 4-tert-butyl pyridine (TBP) molecule case

Luis Antonio Guallichico Guallichico[1], Matheus Josué Souza Matos[1], Mario Sérgio Carvalho Mazzoni
[1]Universidade Federal de Ouro Preto

Simulation of Vavilov-Cherenkov X-rays in GEANT4 near the photo absorption edges of carbon

Dronik Vitaliy[1],Nazhmudinov Ramazan, Kubankin Alexander,Kishin Ivan[1]
[1]Belgorod National Research University

Predicting the Mechanical Properties of Borophene by Machine Learning

José Daniel Moreno González[1]
[1]Escuela Superior Politecnica del Litoral

Synthesis and electrical properties of MoS2

Hillary Estefania Rodriguez Lucas[1], Henrry Marcelo Osorio Calvopiña[1]
[1]Escuela Politecnica Nacional

Observation of X-rays during heating a pyroelectric crystal by an infrared laser

Ramazan Nazhmudinov[1], Kubankin Alexander, Andrei Oleinik, Artemiy Klenin
[1]Belgorod National Research University

Modifications in flat borosilicate glass samples by electron bombardment in the range of 1 to 10 keV

Juan De la Torre[1], Steven Nuñez[1], Karla Estefanía Moyano Quishpe[1], Cristian Patricio Santacruz Teran[1], Isamar Sarabia Ayala[1], Esteban Irribarra[1], Cesar Costa Vera[1]
[1]Escuela Politécnica Nacional

Green synthesis to produce copper nanoparticles using Citrus Meyeri

Joseth Alejandra Pachacama Ruiz[1], Gabriela Viviana Tubon Usca
[1]Escuela Superior Politécnica De Chimborazo

Experimental setup for the study of the interaction of 100 eV – 10 keV electrons with matter

Steven Nuñez[1], José Luis González peñafiel[1], Juan De la Torre[1], Cristian Patricio Santacruz Teran[1], Esteban Irribarra[1], Cesar Costa Vera[1], Ramon Xulvi-Brunet[1]
[1]National Polytechnic School

Transport Properties of Graphene in proximity with Alkali Metals

Cristina Vaca Chanatasig[1]
[1]Yachay Tech University

High-energy physics

A portable particle detector to study natural radioactivity and cosmic rays in open field

Wilson Santiago Reino Cárdenas[1], Marco Schioppa, Davide Passarelli
[1]Università della Calabria

Journal of Physics: Conference Series **2238** (2022) 011001 doi:10.1088/1742-6596/2238/1/011001

Optics

Change of temperature in Current Density Gain in Triple Junction Solar Cells with Different Antireflective Coatings

Darwin Guanga[1], Jaouad Abdelatif, Arnaud Ritou, Maxime Darnon[2], Hamon Gwénaëlle, Volatier Maïté

[1]Sherbrooke University;[2]Centre national de la recherche scientifique

---

Arduino-Phyton interface for a light photometer

Carlos Eduardo Smith Rincon[1], Luis José Borrero González[2], Jesús González-Laprea[2]

[1]Escuela Superior Politécnica del Chimborazo;[2]Pontificia Universidad Católica del Ecuador

---

Development and Implementation of a Universal Quantum Simulator using the Qiskit platform

Gabriel Cofre[1], Jiri Svozilik[2]

[1]Yachay Tech University;[2]Palacký University, Olomouc

Other

how to reduce the pollution produced by motorcycles? : contribution from nanotechnology and materials science

Jose Fernano Muñoz[1], Jorge Enrique Rodríguez Páez
[1]UNIVERSIDAD DEL CAUCA

A comparative study of liquid-crystal organization in healthy and fatty liver

María Isabel Subía Potosí[1], Hernán Andrés Morales-Navarrete[2]
[1]Escuela Politécnica Nacional;[2]University of Konstanz

Hearing health virtual assessment through association rules mining inside a college community

Ruben Alejandro Ortiz Francisco[1], Maria Isabel Subia Potosi, Eliana Maribel Acurio Méndez[1], Hernan Vinicio Barba Molina
[1]Escuela Politécnica Nacional

Effect Of Dehydration On The Chlorophyll A Flourescence Induction: Experiments And Modelling

Erika Quilca[1], Marco Bayas
[1]Escuela Politécnica Nacional

One-step Laser Induced Graphene production for energy storage applications

Kevin Espinoza[1], Luis Fernando Pantoja, Diego Fierro
[1]National Polytechnic School

Conformational Transition of Influenza Hemagglutinin Loop-36 Explored with Molecular Dynamics Simulations

Klever David Cajamarca-Sacta[1], Marco Bayas
[1]National Polytechnic School

Journal of Physics: Conference Series **2238** (2022) 011001 doi:10.1088/1742-6596/2238/1/011001

Dna Strand Breaks Directly Induced By Kev X-Rays

Andres Ordoñez[1], Marco Bayas
[1]National Polytechnic School

Peer Review Statement

All papers published in this volume have been reviewed through processes administered by the Editors. Reviews were conducted by expert referees to the professional and scientific standards expected of a proceedings journal published by IOP Publishing.

- **Type of peer review:** Single Anonymous
- **Conference submission management system:** Morressier
- **Number of submissions received:** 14
- **Number of submissions sent for review:** 13
- **Number of submissions accepted:** 13
- **Acceptance Rate (Submissions Accepted / Submissions Received × 100):** 92.9
- **Average number of reviews per paper:** 2.77
- **Total number of reviewers involved:** 18
- **Contact person for queries:**
  Name: Esteban Irribarra
  Email: esteban.irribarra@epn.edu.ec
  Affiliation: Escuela Politécnica Nacional

Observation of X-rays during heating a pyroelectric crystal by an infrared laser

R M Nazhmudinov[1,2], **A S Kubankin**[1,2], **A N Oleinik**[1,3] **and A A Klenin**[1]

[1] Belgorod National Research University, 85 Pobedy Str., Belgorod 308015, Russia
[2] Lebedev Physical Institute of RAS, 53 Leninsky Ave., Moscow 119991, Russia
[3] Royal Holloway, University of London, Egham Hill, Egham, Surrey, TW20 0EX, Great Britain

E-mail: Ramazan.M.Nazhmudinov@gmail.com

Abstract. A pyroelectric X-ray source is proposed, in which a lithium tantalate crystal is heated by an infrared laser with a wavelength of 10.6 μm. X-ray spectra measured during irradiation of the crystal with infrared radiation and during natural cooling of the crystal include characteristic X-ray radiation of atoms contained in the structural parts of the source, as well as bremsstrahlung of electrons with energies above 50 keV. An 8 mm sodium chloride window was used to inject 64 W infrared radiation into a vacuum chamber with the pyroelectric crystal installed.

1. Introduction

The pyroelectric effect has been known for many years and is used for radiation detection, imaging, radiometry and thermometry, energy conversion, etc. [1, 2]. Pyroelectric materials are spontaneously polarized anisotropic dielectrics, the temperature change of which leads to the formation of an electric potential difference along the polarization axis [3]. The potential difference between the surfaces of a pyroelectric sample placed in a vacuum can reach a value of about 100 kV, and it can be used to accelerate electrons [4] and ions [5], as well as to generate X-rays [6, 7] and neutrons [8, 9]. Active research into the use of pyroelectrics to generate ionizing radiation, described in detail by Geuther [3], prompted Amptek, Inc. [10] to develop a compact, battery-powered pyroelectric X-ray source. Such compact radiation sources can be used for X-ray fluorescence analysis (XRF) [11, 12], X-ray absorption spectral analysis (XAS) [13], X-ray imaging [14], treatment [15], and education [16].

For temperature control of pyroelectrics, simple resistive heaters, Peltier devices (TEC), or liquid nitrogen coolers have usually been used. At the same time, the use of laser radiation [17–19] makes it possible to more accurately control the power supplied to the pyroelectric sample during heating and makes it possible to control the duration of irradiation, providing the possibility of both continuous heating and short pulse exposure. To date, prototypes of X-ray sources have been proposed, in which lithium niobate crystals (LiNbO3), placed in a vacuum, were heated with blue (470 nm) [20] or ultraviolet (266 nm) [21] lasers.

Here we present the X-ray spectra measured by heating a lithium tantalate ($LiTaO_3$) crystal with an infrared laser and by natural cooling.

2. Experiment

The scheme of the experiment performed at Belgorod National Research University (Belgorod, Russia) is shown in figure 1. The lithium tantalate crystal [22] with dimensions of 20 × 20 mm^2 and a thickness of 10 mm was placed in a vacuum chamber, which is a standard KF40 stainless steel cross. An Oerlikon Leybold Turbovac 50 high-vacuum pump was connected to one of the four KF40 flanges through a vacuum valve. The valve was used to set the air pressure in the vacuum chamber to 1×10^{-4} mbar. During the experiment, the pyroelectric crystal glued to a holder made of duralumin (2024 aluminum alloy) using Epotec HD21D electrically conductive epoxy resin was irradiated with infrared radiation with a wavelength of 10.6 μm, which is outside the transmission range for lithium tantalate (approximately 0.4–5.5 μm [23]). The source of infrared radiation was the carbon dioxide (CO_2) laser of an industrial cutting machine. An infrared radiation beam with a power of up to 80 W and a transverse diameter of about 1 mm entered the vacuum chamber through an 8 mm thick sodium chloride (NaCl) window, transparent to radiation with a wavelength of 0.2 to 20 μm [24]. A massive brass target with a 10-mm hole was mounted at 10 mm from the crystal surface.

X-ray radiation generated by a change in the temperature of the pyroelectric crystal escapes the vacuum chamber through a Mylar window 100 μm thick, transparent for radiation with an energy of more than 3 keV [25]. A semiconductor detector Amptek XR100T with cadmium telluride (CdTe) crystal 5 × 5 mm^2 in size and 1 mm thick, installed at 115 mm from the axis of the laser beam, registered the X-ray radiation spectra.

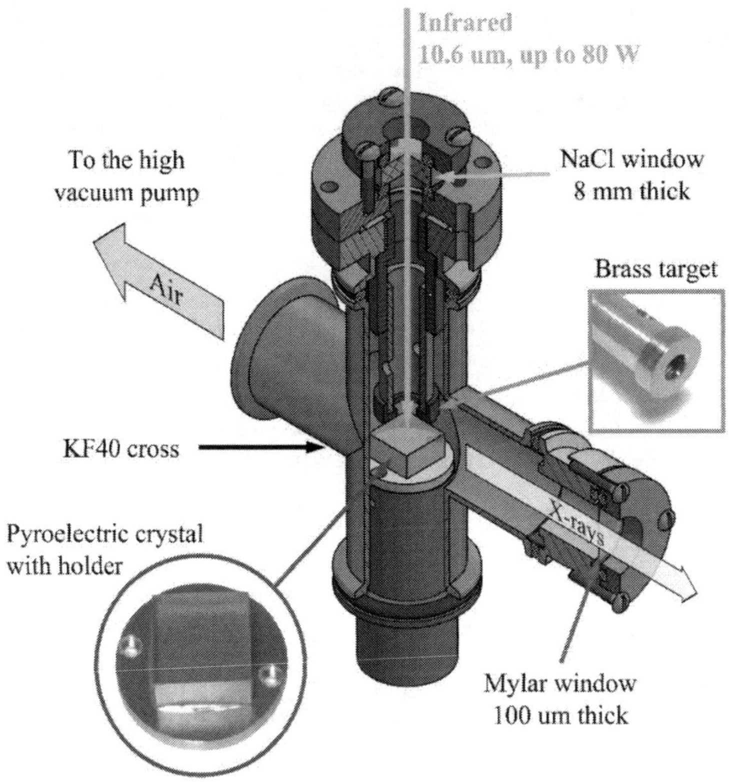

Figure 1. Experimental layout. Infrared radiation enters the evacuated KF40 cross through a sodium chloride window, passes through a hole in the brass target, and hits the lithium tantalate crystal. X-rays produced by changing the temperature of the pyroelectric crystal are output through the Mylar window.

3. Results

During the experiment, infrared radiation with a power of 64 W was generated by the laser, entered through the sodium chloride window into the vacuum chamber, passed through the hole in the brass target, and interacted with the lithium tantalate crystal. Under the action of the beam, the temperature of the crystal increased, and a potential difference was formed between the surfaces of the crystal and the target. This potential difference accelerated the electrons emitted from the pyroelectric surface [17, 26] towards the target, as shown in figure 2(a). When the accelerated electrons hit the target, characteristic X-rays (CXR) [27] and bremsstrahlung are generated. Figure 2(b) shows the X-ray spectrum measured by irradiating the pyroelectric with infrared radiation for 60 seconds. The spectrum contains narrow peaks corresponding to CXR of the copper (Cu) and zinc (Zn) atoms that make up the target, as well as the iron (Fe) atoms contained in the vacuum chamber. The spectrum also contains a wide pedestal with end-point energy of about 53 keV, corresponding to electron bremsstrahlung. The selected radiation power and duration of exposure are dictated by the crystal durability. Exceeding these values (64 W and 60 s, respectively) led to the destruction of the crystal due to stresses caused by rapid inhomogeneous heating.

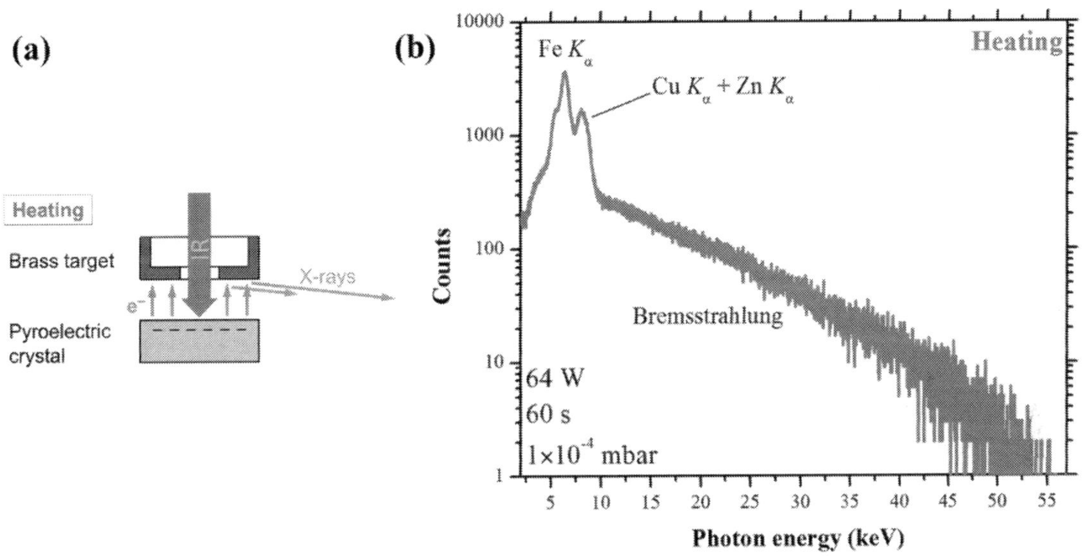

Figure 2. Heating the pyroelectric with the infrared laser (a). When a pyroelectric crystal is heated, an electric field is formed between its surface and a brass target, which accelerates electrons. The X-ray spectrum (b), which is generated during the deceleration of electrons in the target, includes bremsstrahlung and peaks of the characteristic X-ray radiation of target and vacuum chamber atoms.

After turning off the laser, the crystal temperature began to decrease due to natural cooling. Upon cooling, the polarity of the charge formed on the crystal surface changed, and the electrons began to accelerate from the target to the crystal (figure 3(a)). In this case, the X-ray spectrum (figure 3(b)) contains the CXR peaks of tantalum (Ta) atoms contained in the pyroelectric crystal. The low intensity of X-ray radiation and the low value of the end-point energy of bremsstrahlung (about 36 keV) are associated with a low temperature change rate in comparison with the case of heating. A low cooling rate was required to avoid damage to the crystal and increase its lifetime.

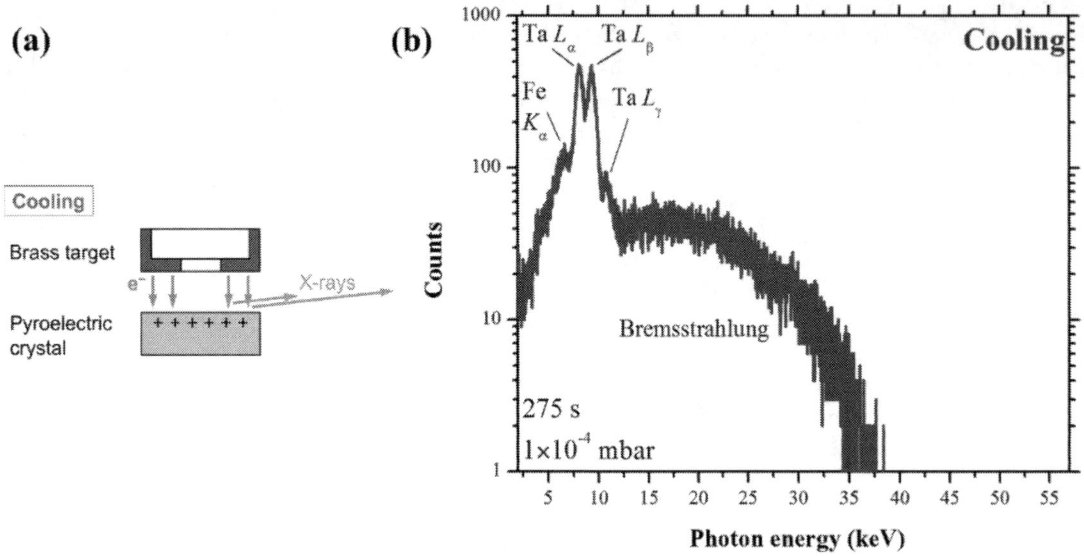

Figure 3. Natural cooling of the pyroelectric (a). When a pyroelectric crystal is cooled, an electric field is formed between its surface and a brass target, which accelerates electrons. The X-ray spectrum (b) generated during the deceleration of electrons in the crystal includes bremsstrahlung and peaks of the characteristic X-ray radiation of crystal and vacuum chamber atoms.

4. Conclusion

We have experimentally demonstrated the possibility of using infrared radiation for heating pyroelectrics in sources of ionizing radiation. The presented radiation source generated X-rays with end-point energy of 53 keV and an intensity of about 6×10^6 photons per second per steradian when a lithium tantalate crystal was heated by a conventional carbon dioxide laser. In addition, lasers provide remote heating, which makes it possible to create a wireless source with an X-ray emitter no larger than a subminiature vacuum tube.

CRediT authorship contribution statement

R M Nazhmudinov: Validation, Formal analysis, Investigation, Writing - Original Draft, Writing - Review & Editing, Visualization; **A S Kubankin:** Conceptualization, Methodology, Resources, Supervision, Project administration; **A N Oleinik:** Resources; **A A Klenin:** Resources.

Acknowledgments

The work was financially supported by a Program of the Ministry of Education and Science of the Russian Federation for higher education establishments, project No. FZWG-2020-0032 (2019-1569). The work of A.O. was supported by a grant from the Russian Science Foundation (project №21-72-00006).

References

[1] Hadni A Applications of the pyroelectric effect 1981 *J. Phys. E: Sci. Instrum.* **14** 1233–40
[2] Joshi J C and Dawar A L Pyroelectric materials, their properties and applications 1982 *Phys. Status Solidi* A **70** 353–69
[3] Geuther J A 2007 *Radiation generation with pyroelectric crystals* (Troy, New York: Rensselaer Polytechnic Institute)

[4] Brownridge J D and Shafroth S M Self-focused electron beams produced by pyroelectric crystals on heating or cooling in dilute gases 2001 *Appl. Phys. Lett.* **79** 3364–6

[5] Geuther J A and Danon Y Electron and positive ion acceleration with pyroelectric crystals 2005 *J. Appl. Phys.* **97** 074109

[6] Brownridge J Pyroelectric X-ray generator 1992 *Nature* **358** 287–8

[7] Geuther J A and Danon Y High-energy x-ray production with pyroelectric crystals 2005 *J. Appl. Phys.* **97** 104916

[8] Naranjo B, Gimzewski J and Putterman S Observation of nuclear fusion driven by a pyroelectric crystal 2005 *Nature* **434** 1115–7

[9] Danon Y Pyroelectric crystal D-D and D-T neutron generators 2012 *J. Instrum.* **7** C04002

[10] https://www.amptek.com/

[11] Ida H and Kawai J Identification of steel by X-ray fluorescence analysis with a pyroelectric X-ray generator 2004 *Anal. Bioanal. Chem.* **379** 735–8

[12] Hatakeyama S, Kunimura S, Sasaki N, Yamamoto T and Kawai J Spatially resolved X-ray fluorescence analysis with a pyroelectric X-ray emitter 2008 *Anal. Sci.* **24** 847–50

[13] Ishii H and Kawai J X-ray absorption spectral analysis with a 9 V battery X-ray generator 2005 *Anal. Sci.* **21** 783-4

[14] Wilke M, Knapp W, Ecke M, Harnisch K, Zierau M and Halle T Investigations of pyroelectric crystals for vacuum electron sources and X-ray applications 2017 *30th International Vacuum Nanoelectronics Conference (IVNC)* pp 304–5

[15] Fullem T Z, Fazel K C, Geuther J A and Danon Y Therapeutic dose from a pyroelectric electron accelerator 2009 *Radiat. Res.* **172** 643–7

[16] Gillich D, Shannon M, Kovanen A, Anderson T, Bright K, Edwards R, Danon Y, Moretti B and Musk J Pyroelectric crystal accelerator in the Department of Physics and Nuclear Engineering at West Point 2011 *AIP Conf. Proc.* **1336** 763–6

[17] Kisa T, Murakami K, Abo S, Wakaya F and Takai M Electron emission from $LiNbO_3$ crystal excited by ultraviolet laser 2010 *J. Vac. Sci. Technol.* B **28** C2B27

[18] Kitamura K, Hatano H, Takekawa S, Schütze D and Aono M Large pyroelectric effect in Fe-doped lithium niobate induced by a high-power short-pulse laser 2010 *Appl. Phys. Lett.* **97** 082903

[19] Louchev O A, Wada S, Saito N, Hatano H and Kitamura K Pulsed laser operated high rate charging of Fe-doped $LiNbO_3$ crystal for electron emission 2012 *J. Appl. Phys.* **112** 073107

[20] Kukhtarev N V, Kukhtareva T V, Stargell G and Wang J C Pyroelectric and photogalvanic crystal accelerators 2009 *J. Appl. Phys.* **106** 014111

[21] Minami K, Ishida T, Asada Y, Shirakawa S, Kobayashi M, Suzuki S, Konishi T and Mimura H Development of a small X-ray source using an ultraviolet laser and pyroelectric crystal 2019 *X-ray Spectrom.* **48** 691–5

[22] Aoyagi S, Osawa H, Sugimoto K, Iwata M, Takeda S, Moriyoshi C and Kurowa Y Crystal structure analysis of $LiTaO_3$ under electric field 2015 *Jpn. J. Appl. Phys.* **54** 10NB03

[23] Loiacono G M Optical transmission variations in lithium tantalate 1968 *Appl. Opt.* **7** 555–6

[24] Kaplunov I, Kropotov G, Rogalin V and Shahmin A Optical properties of alkali halide crystals 2020 *J. Phys.: Conf. Ser.* **1697** 012253

[25] Henke B L, Gullikson E M and Davis J C X-ray interactions: photoabsorption, scattering, transmission, and reflection at E = 50–30,000 eV, Z = 1–92 1993 *At. Data Nucl. Data Tables* **54** 181–342

[26] Rosenman G, Shur D, Krasik Ya E and Dunaevsky A Electron emission from ferroelectrics 2000 *J. Appl. Phys.* **88** 6109

[27] Deslattes R D *et al* 2003 *X-Ray Transition Energies Database, NIST Standard Reference Database Number 128* (National Institute of Standards and Technology Gaithersburg MD 20899) https://dx.doi.org/10.18434/T4859Z

Predicting the mechanical properties of borophene by artificial neural networks

J D Moreno, A López and E D Gutierrez

Escuela Superior Politécnica del Litoral, ESPOL, Campus Gustavo Galindo Km 30.5 Vía Perimetral, Guayaquil, Ecuador

E-mail: jdmoreno@espol.edu.ec, alexlop@espol.edu.ec, egutierr@espol.edu.ec

Abstract. A material of great relevance in the current research context is borophene, a monolayer nanomaterial composed only of boron atoms with extraordinary electrical and mechanical properties. In the present work, a neural network was designed and trained in order to predict the mechanical properties of this material, such as Young's modulus, fracture strength and fracture strain. The training data set was obtained through molecular dynamics simulations, with different parameter scenarios in order to analyze the effects of temperature, strain rate and strain direction. The trained machine learning model succeeded in predicting the material's behavior with high accuracy. Its results reflect a decrease in yield stress with increasing temperature and a slight improvement in the fracture strain with increasing strain rates, as expected. Additionally, a web application with a graphical interface was developed, which uses the trained model, in order to make this tool available to any user. This interface also makes it possible to visualize the approximate stress-strain curve, drawn based on the resulting fracture stress and strain.

1. Introduction

Following the great discoveries made in the past years on graphene's properties, such as its high electrical and thermal conductivity, flexibility, and its possible applications in the field of sensors [1] and energy storage [2], researchers are interested in similar 2D materials and their potential applications [3–5]. A promising material increasingly studied in the past few years is borophene, a monolayer nanomaterial composed only of boron atoms. It was first discovered theoretically in the 1990s [6] and synthesized in recent experiments, using techniques such as thermal evaporation deposition [7], chemical vapor deposition [8], liquid-phase exfoliation or molecular beam epitaxy methods [9], usually on a silver substrate under ultrahigh vacuum.

Borophene can be found in different crystal structures depending on the synthesis process. The main phases that have been discovered are the triangular, β_{12} and χ_3 (displayed in Figure 1). In the 3D triangular phase structure, the atoms are categorized into top chain and bottom chain, separated by a buckling height h of 0.91 Å [10], while the other phases are planar.

Studies have shown that borophene is highly anisotropic [10], meaning that its properties will vary depending on the direction they are measured. The two in-plane directions are known as the zigzag and armchair directions, named after the shape of the sheet's edges (Figure 1(a)).

Borophene was also found to be a very flexible and elastic material [11] and it is predicted as the lightest 2D metal [12], with a really low density in comparison with similar nanomaterials [13]. Studies have also shown a high fracture strength. In its β_{12} phase, it can reach up to 14

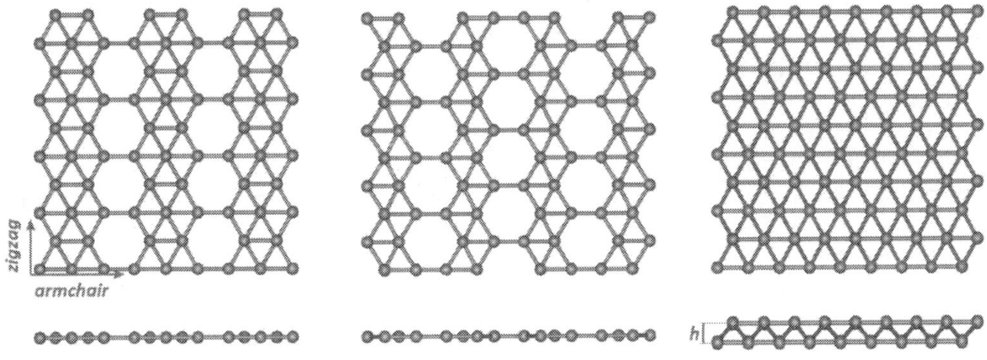

Figure 1. Top and front views of triangular crystal structure of borophene. The 2D sheets (a) β_{12} and (b) χ_3. Both β_{12} and χ_3 boron sheets are planar without vertical buckling but with different arrangements of periodic holes. In (c) a triangular phase with a buckling height h. Figures taken from [16].

N/m in the zigzag direction and 15 N/m in the armchair direction, with critical strains in the range of 10 to 15% at 0 K [14], while in its triangular phase, the values of fracture strength can reach up to 12.3 N/m and 32.9 N/m for the zigzag and armchair directions respectively, with fracture strains of 17% and 16%, at 1 K [15]. A high Young's modulus has also been reported for this material. In its triangular phase it can reach values of 398 N/m for the armchair direction and 163 N/m for the zigzag direction [14]. For comparison, graphene´s modulus can reach values up to 338.08 N/m [10]. As a consequence of its particular structure, the triangular phase of borophene presents a bending modulus, or the tendency to resist bending, of 4.9 eV along the armchair direction [14], and a negative Poisson's ratio in the out-of-plane direction. This means that the width of the borophene sheet will increase, proportional to the magnitude of the Poisson's ratio, when strained along a particular direction. The values found for this coefficient are -0.04 when deformed along the armchair direction and -0.02 when deformed along the zigzag direction [8]. First principle calculations have also shown important electronic, optical and thermodynamic properties [17], with a great potential as anode material for lithium, sodium and magnesium ion batteries [10]. It has also shown an intrinsic superconducting behavior, with critical temperature estimated in the range of 10 to 20 K [14].

Although great strides have been made to determine the characteristics and properties of borophene, there is still much work to be done. So far, the information available in the literature comes predominantly from first-principles calculations and molecular dynamics (MD) simulations. These methods involve a considerable processing complexity and, therefore, a lot of time and resources, which makes the information available very limited. In addition, a new simulation or calculation must be carried out after each experiment. This in turn implies varying the input parameters or constraints, which ends up slowing down the research process. That is why, this work proposes to use a model created from a neural network, based on similar previous studies carried out with graphene [18]. This model will allow predicting the mechanical properties of triangular borophene such as Young's modulus, fracture stress and maximum deformation in a matter of seconds, which will make it possible to repeat the experiment under different conditions without the need for additional simulations or calculations.

The paper is structured as follows. Section 2 deals with the methods used to obtain the data set, describes the software and configurations used in the molecular dynamics experiments. In addition, data on the interatomic potential used and its adaptation for borophene are provided. The operation of artificial neural networks and the architecture used for the predictive model's design are also described. Section 3 presents the results on the behavior of the material under different control parameters, like temperature, strain rate and strain direction, obtained from the simulations. A negative correlation between temperature and Young's modulus, fracture strain and fracture strength is discussed. This section also presents the results from the trained model and the errors associated with each predicted variable. The last section summarizes the information obtained and how it can contribute to the development of research in this and other materials.

2. Methods

2.1. MD simulations and data gathering

As an initial step, a dataset was collected from molecular dynamics simulations carried out in LAMMPS (Large-scale Atomic/Molecular Massively Parallel Simulator) [19], an open-source software that allows modeling sets of particles, from atomic systems to more complex arrangements such as polymers, metallic crystals and other materials in solid, liquid or gaseous state. In this software, a square sheet of borophene of 60 Å each side, corresponding to 1440 boron atoms, was simulated. In these simulations, the Stillinger Weber (SW) potential parameterized by Jiang *et al.*[15] was used to model the interactions between atoms.

This potential consists of a two-body term and a three-body term. The two-body term represents the stretching of the bond, and is given by

$$V_2(r_{ij}) = \epsilon A(B\sigma^4 r_{ij}^{-4} - 1)e^{[\sigma/(r_{ij}-r_{max})]} \quad (1)$$

where r_{ij} is the length of the bond between atoms i and j, and r_{max} is the cut-off distance. The three-body term corresponds to the distortion angle and is given by

$$V_3(\theta_{ijk}) = \epsilon\lambda e^{[\gamma\sigma/(r_{ij}-r_{max})+\gamma\sigma/(r_{jk}-r_{max})]}(cos\theta_{ijk} - cos\theta_0)^2 \quad (2)$$

where θ_{ijk} is the angle between bond ij and bond jk. The geometric parameters σ, B, γ and θ_0, and the energy parameters A, ϵ and λ, used in the SW potential for borophene are summarized in Table 1.

Table 1. Parameters used in equations 1 and 2 of the SW potential, used in LAMMPS. The two-body and three-body interactions found in the system, can be summarized in 4 main types of bond, shown in the first column.

Bonds	ϵ (eV)	σ (Å)	λ	γ	cosθ_0	A	B
$B_1 - B_5 - B_5$	1.000	0.971	0.000	1.000	0.000	8.974	3.817
$B_1 - B_2 - B_2$	1.000	0.618	0.000	1.000	0.000	2.098	42.820
$B_1 - B_2 - B_5$	1.000	0.000	32.074	1.000	0.429	0.000	0.000
$B_1 - B_2 - B_4$	1.000	0.000	23.668	1.000	-0.162	0.000	0.000

Given that each boron atom has 6 neighbours, and the angles that it forms with some of them are not distinguishable in LAMMPS, this parameterization [15] proposes a unit mesh made of 12 atoms, each one of them assigned with an atom type (Figure 2). Before the deformation experiments, the borophene sheets were equilibrated using the isothermal-isobaric (NPT) ensemble for 50 *ps*. For the strain test simulations, periodic boundary conditions were used in all 3 axes, and uniaxial displacement was applied to the boundary corresponding to

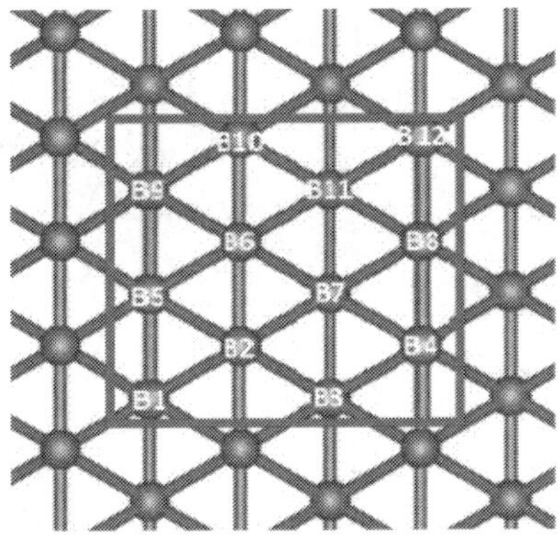

Figure 2. Unit mesh of the borophene's triangular structure used in the parameterized SW potential, with the corresponding atom types from B1 to B12.

either the zigzag or armchair direction, with a time step of 0.1 fs. Several uniaxial deformation test scenarios were simulated until the fracture of the material. These scenarios consist of temperatures of 1, 10, 100, 200, 300 and 500 K; strain rates of 0.00025, 0.0005, 0.00075, 0.001, 0.0025 and 0.005 ps^{-1}; five initial system speed conditions, and the two deformation directions, giving a total of 360 simulations, with each of them taking from 400,000 to 8,000,000 steps depending on the parameters used. The five initial velocity ensembles were randomly generated using five different seeds chosen arbitrarily. In this way, every parameter configuration will be simulated five times, with different initial speed conditions each time. The seed values are: 12345, 12346, 12347, 12348 and 12349.

The upper range of temperatures was chosen based on the critical value of borophene, which is equal to 550 ± 50 K [20]. Above this temperature, the borophene structure becomes unstable, and therefore, not suitable for the stress tests used in this work. It was also taken into account to include the range from 10 K to 20 K, within which borophene has superconducting properties [14]. The range of strain rates was defined considering the time it would take for the simulation of greater duration. The simulations at the highest strain rate chosen (0.005 ps^{-1}) had an approximate duration of 12 minutes each, while each simulation at the lowest strain rate (0.00025 ps^{-1}) took about 4 hours. All the 360 simulations corresponded to about 500 hours. The results are then gathered into a table with the feature values of temperature, strain rate and direction of each simulation with its corresponding Young's modulus, fracture stress and maximum strain. Subsequently, this data set was expanded using a synthetic data generation algorithm based on a Gaussian Copula model provided by the SDV python library [21]. This model was modified in such a way that it takes the five simulation results corresponding to the five initial speed conditions of a given parameter configuration as input, and outputs 15 new examples based on the behaviour of the input values. The process is repeated for every parameter configuration, giving a total of 1080 new examples. As a result, the dataset now has 1440 examples.

2.2. Artificial Neural Networks
Artificial neural networks (ANN) are computational models that seek to mimic the way in which neurons in our brain learn [22]. Its structure is organized in layers, each one made up of one or more processing units (figure 3). A neural network will have an input layer, one or more hidden or intermediate layers and an output layer. The input values are propagated from layer to layer, while each of the processing units performs a computation of the data, based on the gradient descent algorithm [23]. This process is repeated back and forth through the network, in what is called the error backpropagation [24]. In this way, the unit's predictions are compared with the

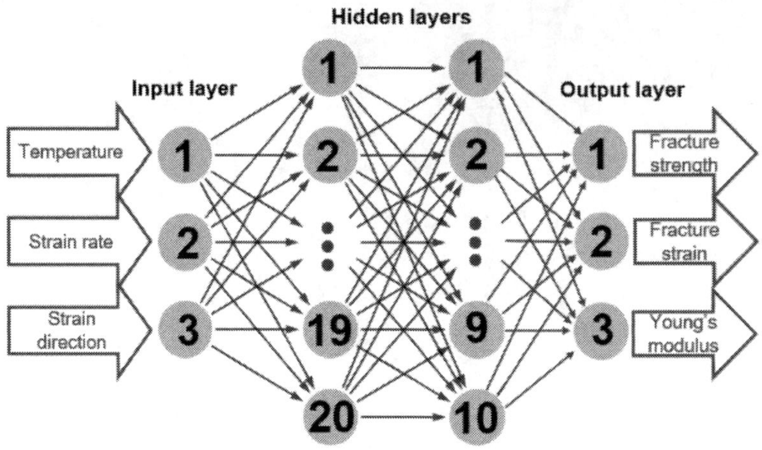

Figure 3. Structure of the neural network designed for the model.

real values in terms of the mean squared error (MSE) with the goal of reducing it. For the design of the neural network, the Keras library was used, and a 4-layer architecture was chosen (Figure 3). There is an input layer with three units, one for each input parameter (direction, strain rate and temperature), followed by two intermediate layers, with 20 and 10 units respectively and an output layer with 3 units, one for each value to be predicted (Young's modulus, fracture stress and maximum strain).

For optimal training of a neural network, it is necessary to divide the input data set into three subsets: training, validation and test set. The training set is the largest of the three. It is reserved for training the model. The validation set is used alongside the training of the model to get an idea of how well it is responding to new data that was not originally used during the training step. This is done in order to avoid the model to be excessively fitted to the training data. Depending on how the neural network responds to this new data, different parameters of the model will be adjusted, through trial and error. In this way, a configuration is obtained that favors a better performance of the model over the validation set. Finally, to evaluate the overall performance of the resulting model, it is necessary to have a totally new data set that is not related to the parameter adjustment process. This is known as the test set. For the borophene dataset, 20% of the data was separated for the test set, 20% for the validation set and the remaining 60% for the training set, corresponding to 288 examples for testing, 288 for validation, and 864 for training.

3. Results

To check the veracity of the data obtained from the simulations carried out, the results of the simulations at 1 K and 300 K at a strain rate of 0.001 ps^{-1} were taken and compared with the results of the work of Jiang *et al.* [15]. Figure 4 shows the comparison of both works through their stress *vs.* strain curves, where a similar behavior can be seen for each of the curves. In both figures the fracture point reaches higher values of stress and strain along the armchair direction. For a temperature of 300 K the fracture strength values for the armchair and zigzag direction are 26.1 N/m and 10.58 N/m respectively, and the maximum strains are equal to 9.3% and 11%. At 1 K, the corresponding fracture stress values for the armchair and zigzag directions are 32.94 N/m and 12.35 N/m, with a maximum deformation of 15.7% and 17.1% respectively. Young's modulus was obtained from a linear fit in each resulting curve, in the range of 0 to 1% following [15]. The values obtained were 160.98 N/m and 152.92 N/m for the deformation along

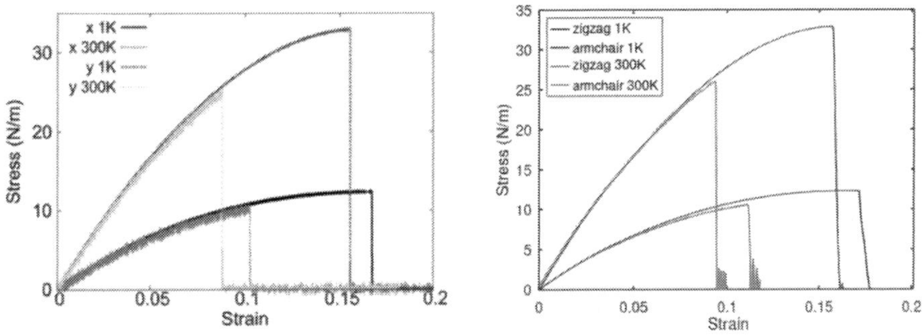

(a) (b)

Figure 4. Comparison between stress vs strain curves (a) from Jiang *et al.* [15] and (b) from this work, for temperatures of 1 K and 300 K. Values in (b) were filtered to reduce noise.

the zigzag direction, at 1 K and 300 K, respectively. For the deformation along the armchair direction, values of 381.18 N/m and 370.73 N/m were obtained, respectively. These data were compared with other sources, their results are summarized in Table 2.

Table 2. Comparison of the mechanical properties of borophene from different sources.

	Young's modulus (N/m)	**Fracture strength (N/m)**	**Fracture strain (%)**	**Temp.**	**Reference**	**Method**
Zigzag direction	162.7	12.3	17.0	1K	Zhou et al. [15]	MD simulations
	163	13.2	15.3	1K	Zhou et al. [20]	MD simulations
	165.03	12.32	16.7	1K	Pham et al. [25]	MD simulations
	161.74	12.48	16.68	1K	This work	Predictive model
	160.98	12.35	17.1	1K	This work	MD simulations
	152.92	10.58	11	300K	This work	MD simulations
	153.19	10.45	10.62	300K	This work	Predictive model
	150.77	11.23	10.2	300K	Zhen et al. [26]	Ab initio
Armchair direction	385.0	32.9	16.0	1K	Zhou et al. [15]	MD simulations
	394	24.1	10.2	1K	Zhou et al. [20]	MD simulations
	390.02	32.88	15.6	1K	Pham et al. [25]	MD simulations
	380.12	32.8	15.5	1K	This work	Predictive model
	381.18	32.94	15.7	1K	This work	MD simulations
	370.73	26.1	9.3	300K	This work	MD simulations
	369.12	25.08	8.8	300K	This work	Predictive model
	357.88	26.83	8.7	300K	Zhen et al. [26]	Ab initio

Table 2 shows slight discrepancies between this work's MD results and other reported simulation results [25], [15]. This is mainly due to the different configurations with which the experiments in each study were made. To begin with, the time step used in the present work (0.0001 *ps*) is lower than the ones in previous studies, in which they both use 0.001 *ps*. This was done in order to achieve a better resolution on the output data. The borophene sheet's dimensions also varied in each study. Zhou *et al.* [15] simulated a 100 × 100 Å sheet, and Pham *et al.* [25] used a 110 × 100 Å sheet, which are larger than the 60 × 60 Å sheet used in this study, and thus imply a greater number of atoms to simulate. Besides, the values reported by both consulted sources were obtained using a fixed strain rate of 0.001 ps^{-1} while the present study used five different values, and the time used to equilibrate the sheets prior to their deformation (100 *ps*) was higher than the one used in this work (50 *ps*). These modifications were made in order to reduce the computational cost, which was a necessary step given the high number of simulations required to gather a training dataset.

3.1. Model training and performance

The feature values of the 864 training examples were normalized in the range of 0 to 1 to have them all within the same range. All units of the neural network were set to use the ReLU function as the activation function. This is a linear function that returns the input directly if it is positive, otherwise it returns 0. It is typically used in regression problems, like the one in the present work. The training was carried out with 150 iterations or epochs, using the Adam optimization algorithm [27] with a learning rate value of 0.01. Once trained, the neural network was used to predict the mechanical properties of borophene. The accuracy of the results is shown in figures 5(a), 5(b) and 5(c), for the fracture strain, fracture strength and Young's modulus respectively. The horizontal axis in the three figures shows the real values taken from the original dataset, and the vertical axis displays the respective model's predictions for each one of the original examples. The figures show how the points predicted by the model fit the diagonals in black, which serve as a reference to know the position of the exact values on where the points of the predictions should fall for each example. To better understand the performance of the model, the error values in terms of the Root mean squared error (RMSE) are shown in Table 3.

Table 3. Error (RMSE) of the three predicted properties.

Fracture strain (%)	Fracture strength (N/m)	Young's modulus (N/m)
0.19	0.29	2.48

3.2. Effects of temperature and strain rate

To study the behaviour of the predicted data as a whole, a new dataset was generated using the trained model, with the same parameter configurations of the MD simulations as input, adding new temperatures in between the existing ones. Figure 6 compares the incidence of temperature on fracture strength, Young's modulus and fracture strain for the MD results and the newly generated dataset. It can be evidenced that, in general, temperature has a negative effect on these properties as expected [20, 25], with similar sensitivities both in the armchair and zigzag directions. The predicted values show that, as the temperature increases from 0 to 500 K, the fracture strength, fracture strain and Young's modulus drop by 28%, 50% and 8% respectively for the zigzag direction. In the armchair direction the values of these properties drop by 31%, 50% and 5% respectively, with the same increase in temperature. Variations in the strain rate do not significantly influence the fracture strength and Young's modulus, however, an increase on the strain rate has a positive impact on the fracture strain [20, 25]. This is

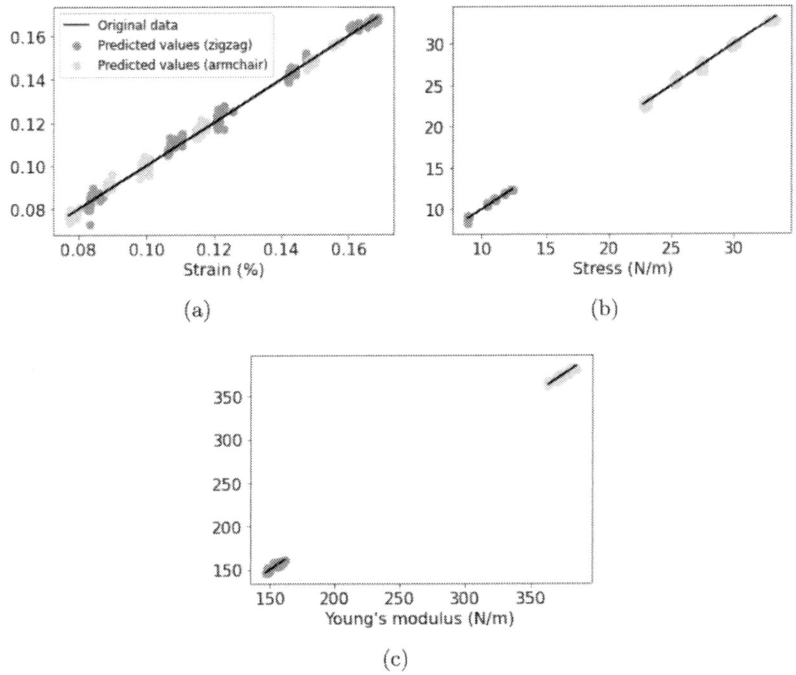

Figure 5. Training results for (a) the fracture strain, (b) the fracture strength and (c) the Young's modulus.

reflected in the predicted values, where fracture strain increases by 2% and 4% in the armchair and zigzag directions respectively, as the strain rate increases from 0.0001 to 0.005 ps^{-1} at room temperature. This can be observed in Figure 7, for all the temperatures and in both strain directions. It can be seen that the sensitivity to variations in the strain rate increases at higher temperatures for both strain directions.

3.3. Web application

A web application with a graphical interface was designed in ReactJS to work alongside the trained model. Its backend was designed using Python's Flask framework. Figure 8(a) shows the graphical interface. It has a box to select the options corresponding to the material and the strain direction, and two sliders to choose the values of strain rate and temperature. In this way, it is ensured that the parameters entered are within the options and ranges allowed for the tool. The box on the right displays the resulting values of fracture stress, maximum strain and Young's modulus. This tool also presents a section that shows the stress strain curve corresponding to the calculation made in the previous section (Figure 8(b)). The fracture point coincides with the predicted results. This curve is approximate to the results of the calculation, but not exact. This is due to the fact that the data extracted from the simulations, used for training the model, considers the stress and strain values only at the instant of fracture. That is why it was necessary to create a model that allows describing these curves only with the information provided by the fracture stress and strain.

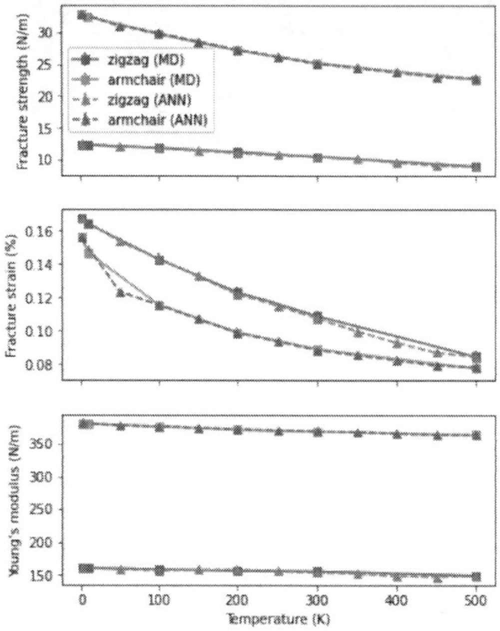

Figure 6. Comparison between MD simulations and the ANN predictions of the mechanical properties of borophene through different temperatures.

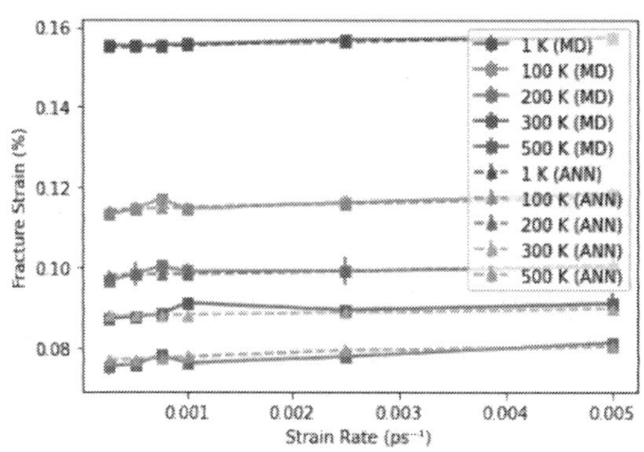

Figure 7. Comparison between MD simulations and the ANN predictions of the fracture strain values for different strain rates and temperatures in the zigzag direction.

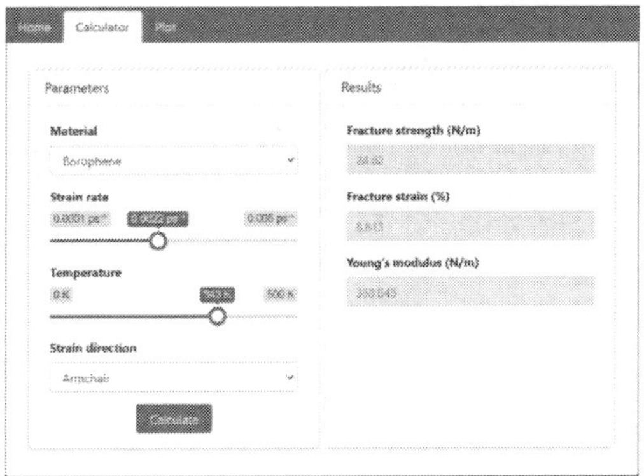

(a)

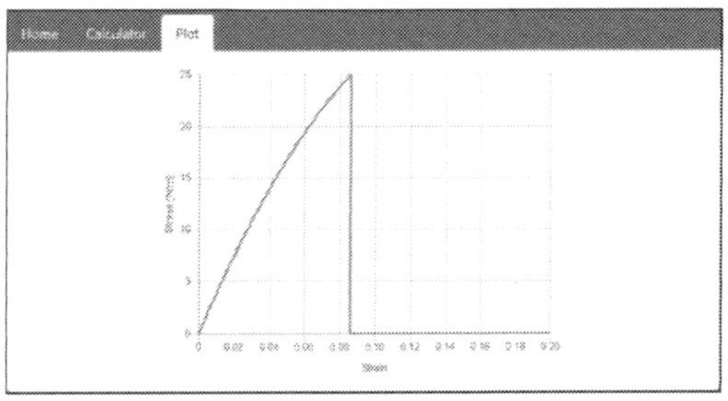

(b)

Figure 8. (a) Graphical interface of the web application. An example is shown of the predictions at 343 K and a strain rate of 0.0022 ps^{-1}. (b) The resulting stress-strain curve of the previous calculation.

4. Conclusion

A neural network was designed in this work to predict the mechanical properties of borophene with promising results. The model was trained with a dataset obtained from molecular dynamics simulations, from which the impacts of temperature, strain rate and strain direction on the material's fracture strength, fracture strain and Young's modulus were analyzed in order to validate them against previous studies. The predicted values showed the expected negative correlation between the target properties and the temperature of the material, as well as an slight improve on the fracture strain, with higher values of strain rate. This predictions are in good agreement with the MD simulation results and Ab initio calculations made by previous studies. In addition, a web application was designed as a graphical user's interface to make use of the trained model, and it allows to visualize an approximation of the material's stress-strain curve drawn by a model based on the predicted fracture stress and strain values. This work serves as an example of the suitability of the use of machine learning models in material science. It can significantly accelerate discoveries on materials like borophene and many others, since these computational methods can be applied to a great variety of properties on many kinds of materials, even before being synthesized.

References

[1] Liu Q, Chen J, Li Y and Shi G 2016 *ACS Nano* **10** 7901–7906 ISSN 1936-0851
[2] Yoo J J, Balakrishnan K, Huang J, Meunier V, Sumpter B G, Srivastava A, Conway M, Mohana Reddy A L, Yu J, Vajtai R and Ajayan P M 2011 *Nano Letters* **11** 1423–1427 ISSN 1530-6984
[3] You Y, Yang C, Zhang X, Lin H and Shi J 2021 *Materials Today Nano* **16** 100132 ISSN 2588-8420
[4] Akhtar M, Anderson G, Zhao R, Alruqi A, Mroczkowska J E, Sumanasekera G and Jasinski J B 2017 *npj 2D Materials and Applications* **1** 5 ISSN 2397-7132
[5] Arcudia J, Kempt R, Cifuentes-Quintal M E, Heine T and Merino G 2020 *Phys. Rev. Lett.* **125**(19) 196401
[6] Boustani I 1997 *Surface Science* **370** 355–363 ISSN 0039-6028
[7] Feng B, Zhang J, Zhong Q, Li W, Li S, Li H, Cheng P, Meng S, Chen L and Wu K 2016 *Nat Chem* **8** 563–568
[8] Mannix A J, Zhou X F, Kiraly B, Wood J D, Alducin D, Myers B D, Liu X, Fisher B L, Santiago U, Guest J R, Yacaman M J, Ponce A, Oganov A R, Hersam M C and Guisinger N P 2015 *Science* **350** 1513–1516
[9] Bhattacharyya G, Mahata A, Choudhuri I and Pathak B 2017 *Journal of Physics D: Applied Physics* **50**
[10] Wang Z, Lü T Y, Wang H Q, Feng Y P and Zheng J C 2016 *Phys. Chem. Chem. Phys.* **18**(46) 31424–31430
[11] Zhang Z, Yang Y, Penev E S and Yakobson B I 2017 *Advanced Functional Materials* **27** 1605059
[12] Mannix A J, Zhang Z, Guisinger N P, Yakobson B I and Hersam M C 2018 *Nature Nanotechnology* **13** 444–450 ISSN 1748-3395
[13] Ranjan P, Sahu T K, Bhushan R, Yamijala S S, Late D J, Kumar P and Vinu A 2019 *Adv Mater* **31** e1900353
[14] Zhang Z, Penev E S and Yakobson B I 2017 *Chem. Soc. Rev.* **46**(22) 6746–6763
[15] Jiang J W and Zhou Y P 2017 *Handbook of Stillinger-Weber Potential Parameters for Two-Dimensional Atomic Crystals*
[16] Peng B, Zhang H, Shao H, Ning Z, Xu Y, Ni G, Lu H, Zhang D W and Zhu H 2017 *Materials Research Letters* **5** 399–407 (*Preprint* `https://doi.org/10.1080/21663831.2017.1298539`)
[17] Peng B, Zhang H, Shao H, Xu Y, Zhang R and Zhu H 2016 *J. Mater. Chem. C* **4**(16) 3592–3598
[18] Zhang Z, Hong Y, Hou B, Zhang Z, Negahban M and Zhang J 2019 *Carbon* **148** 115–123 ISSN 0008-6223
[19] Plimpton S 1995 *Journal of Computational Physics* **117** 1–19 ISSN 0021-9991
[20] Zhou Y P and Jiang J W 2017 *Scientific Reports* **7** 45516
[21] Patki N, Wedge R and Veeramachaneni K 2016 The synthetic data vault *2016 IEEE International Conference on Data Science and Advanced Analytics (DSAA)* pp 399–410
[22] Grossi E and Buscema M 2008 *European journal of gastroenterology and hepatology* **19** 1046–54
[23] Ruder S 2016 *CoRR* **abs/1609.04747** (*Preprint* `1609.04747`) URL `http://arxiv.org/abs/1609.04747`
[24] Rumelhart D E, Hinton G E and Williams R J 1986 *Nature* **323** 533–536 ISSN 1476-4687
[25] Pham V T and Fang T H 2020 *Thin Solid Films* **709** 138197 ISSN 0040-6090
[26] Sha Z D, Pei Q X, Zhou K, Dong Z and Zhang Y W 2018 *Extreme Mechanics Letters* **19** 39–45 ISSN 2352-4316
[27] Kingma D and Ba J 2014 *International Conference on Learning Representations*

Transport properties of graphene in proximity with alkali metals

M Peralta[1], C Vaca-Chanatasig[1], R Vera-Nieto[1] and D Verrilli[2]

[1] Yachay Tech University, School of Physical Sciences and Nanotechnology, 100119 Urcuquí, Ecuador

[2] Centro de Física Teórica y Computacional, Escuela de Física, Facultad de Ciencias, Universidad Central de Venezuela, Caracas 1050, Venezuela

E-mail: mperalta@yachaytech.edu.ec

Abstract. In this work, we propose the analysis of the electronic and transport properties of graphene decorated with Lithium and Potassium adatoms. We will study two inequivalent metal adsorption sites: the Top site, on top of a carbon atom of one sub-lattice of graphene; and the Hollow site, in the middle of a C6-unit. With this end, we will use an analytical Tight Binding Model, for graphene with adsorbate atoms of lithium and potassium, for the two different adsorption positions. Then, we use the Green's function equation of motion method to calculate the corresponding band structures and density of states, and numerical calculations for the conductance are performed with the quantum transport simulation package of python (Kwant). We find that the bands are down shifted with respect to pristine graphene, indicating a doping with electrons. For the Top case, the AB symmetry breaking produced in this configuration, generates small bandgaps of approximately 170 meV for potassium and 220 meV for lithium. Finally, the conductance is shifted in energy in the same way as the bands, preserving its growing rate with the absolute value of the energy as for pristine graphene.

1. Introduction

In the recent years it has been a huge amount of works about the enhancement of the properties of graphene decorated with adatoms[1, 2, 3, 4, 5, 6, 7, 8]. For example, the adsorption of transition metals can induce ferromagnetic order and high spin orbit coupling in graphene [9, 10, 11, 12]. Also, the adsorption of I–III elements, Pd, Au, and Sn, can distort the graphene sheet, changing its distinctive sp^2-like orbitals to sp^3 ones which are more covalent reactive [1].

A case of high technological interest, is the one of alkali metals such as Lithium (Li) and Potassium (K) on graphene, because of the potential applications in energy storage[13, 14]. Besides, it has been found evidence of superconductivity in Li on graphene[15, 16]. K as well as Li, is an important adatom since it allows the study of graphene quasiparticle dynamics and minimum conductivity[1]. In addition, computational studies of K adsorbate on graphene had demonstrated a long range interaction of the K atom with the graphene surface, revealing a large dipole moment[17].

The stable site of adsorption of Li and K on graphene has been reported as the hollow site, in which the adatoms are placed in the center of the hexagons of the graphene lattice [18, 19, 20]. In this configuration there is a higher coordination and lower charge density, so they can be accommodated closer to the surface, in comparison with other Li(K)-graphene adsorption structures[20]. Accordingly, the band structure and in general, the electronic and transport properties of Li and K on graphene, in the hollow adsorption site, has been modeled using, mainly, computational methods, such as Density Functional Theory (DFT) [20, 21, 22, 23, 24, 25, 26, 27, 28, 29]. Analytical methods based on tight-binding are valuable to complement computational studies of the described structures, because they allow to understand the origin and the form of the interactions in the system, so they can be controlled and analyzed in detail. For example, they can be turned on and off depending on the positions of the adatoms, they can be modulated by changing the distance to the adatoms to investigate the effect on the transport properties, also other interactions like spin orbit coupling and/or magnetism can be added depending on the nature of the adatoms.

Motivated mainly by the few theoretical studies based on analytical methods [30, 31, 32], and by the absence of studies for Li and K on graphene in other non-equilibrium configurations, we present in this work a theoretical model, based on effective parameters tight-binding (TB) Hamiltonians for the calculation of the band structures and densities of states of graphene decorated with Li and K adatoms in two configurations: i) Top (T), in which the adatoms are placed on top of the atoms of the graphene A sublattice, and ii) Hollow (H), where the adatoms are in the center of every graphene hexagon. We find that, as reported in the literature[22, 27], the graphene π bands are down shifted by the presence of the adatoms, indicating a doping of the graphene with electrons. An interesting feature observed in the T case is the opening of a small gap of around 170 meV for K and 220 meV for Li. The TB Hamiltonian presented is then used to calculate the conductance of graphene perturbed by the adatoms by using the quantum transport simulation package of python (Kwant), finding that it grows linearly with the absolute value of the energy, almost at the same rate as pristine graphene.

In section 2, we describe the TB Hamiltonians in the direct space for Li and K on graphene in the two configurations mentioned and their respective transformations to the Fourier space. Then, we show the Green's functions calculations by the equation of motion method in section 3. This method offers the possibility of calculating the bands and the densities of states, avoiding the diagonalization of the Hamiltonians. The results of the quantum conductance of graphene decorated with Li and K are presented in section 4. Finally the conclusions and final remarks are presented in section 5.

2. Graphene with Alkali metals adsorbed

The configurations studied in this work, for Li and K adsorbed on graphene, are presented in Fig. 1, where we have two kinds of staking, namely a top (T) in which the adatom is on top of every atom of the graphene A sublattice (see Fig. 1a), and a hollow (H) for the case in which the adatoms are in the center of the hexagonal ring C_6 (see Fig. 1b). The vectors between graphene atoms A and B at nearest neighbors, $\boldsymbol{\delta}_n$, at second nearest neighbors, $\boldsymbol{\Delta}_j$, and between the graphene atoms A and B and the adatoms denoted by the index l, $\tilde{\mathbf{n}}^A_{i,l}$ and $\tilde{\mathbf{n}}^B_{j,l}$ respectively, are all defined in Fig. 1.

For all our calculations we consider a concentration of two carbon atoms for every adatom (this is $Li(K)C_2$). Besides, we are interested in a situation in which the adatoms are at a vertical distance d such that the graphene lattice is not deformed by their proximity, therefore the distance between carbon atoms is the carbon-carbon bond length $a_{CC} = a/\sqrt(3) = 1.42$ Å, where a is the pristine graphene lattice constant (the distance between unit cells). In Table 1 are shown the equilibrium distances between the graphene plane and the adatoms for the T and H configurations according with DFT first principles calculations [20, 27, 29]. The parameter $b \equiv \sin\theta$ is also included in this table, being θ the angle with the $\hat{z}$ axis of the vector that goes from A(B) to the adatom (see Fig. 1).

A tight binding Hamiltonian, based on effective parameters, can be written for each of these two stacking types (the detailed derivation of the Hamiltonians presented here will be shown elsewhere), T and H, taking into account only the kinetic and the atomic potential as the predominant interactions in this system. The adatoms, are considered as a perturbation of the graphene lattice, modifying the hopping between the p_z orbitals at A and B graphene's sublattices and the onsite energies of electrons in these orbitals. We find, that depending on the position of the adatom, next nearest neighbors hopping terms will contribute at the same order as the perturbative nearest neighbors terms, this can be seen for example, in Fig. 1a, where the hopping for going from B (denoted by i) to the adatom (denoted by l) and then back to the same B, contributes to the same order than the one for going from B to the adatom and then to another second neighboring atom B (denoted by j). All these possibilities of hopping, together with their dependence with the location of the atoms, are considered to write the TB Hamiltonians for Li and K in the T and H configurations.

Table 1: Equilibrium distances d for Li and K on graphene in the top (T) and hollow (H) configurations. These distances are taken from reference[29], which takes into account the Van der Waals corrections in these structures. The parameter $b \equiv \sin\theta$ is given in the fourth column, where θ is the angle of the A(B)-adatom vector with the $\hat{z}$ axis.

	Site	d (Å) [29]	$b \equiv \sin\theta$
Li	Top (T)	2.043	0.57
	Hollow (H)	1.776	0.62
K	Top (T)	2.891	0.44
	Hollow (H)	2.780	0.45

2.1. TOP Configuration

The tight binding Hamiltonian for the T configuration can be written in terms of the creation and annihilation operators for electrons on a site in the nth primitive cell of the A and B graphene sublattices,

$$H_{\rm T} = -(\gamma_0 + t_0)\sum_{\langle ij\rangle}(a_i^\dagger b_j + b_j^\dagger a_i) - t_{1a}\sum_i a_i^\dagger a_i - \Big[\sum_i b_i^\dagger b_i (t_{1b} + t_{2b}(\tilde{\boldsymbol{n}}_{il}^B \cdot \tilde{\boldsymbol{n}}_{il}^B)) + \sum_{\langle\langle i,j\rangle\rangle}(b_i^\dagger b_j + b_j^\dagger b_i)(t_{1b} + t_{2b}(\tilde{\boldsymbol{n}}_{il}^B \cdot \tilde{\boldsymbol{n}}_{jl}^B))\Big], \quad (1)$$

where the first term is the kinetic term, with $\gamma_0 = 3.03\,\text{eV}$ is the hopping parameter between p_z orbitals for pristine graphene, and t_0 is the perturbation to this hopping that appears because now electrons can go from A to the adatom and then to a nearest neighbor B. The second term is the onsite energy of p_z orbitals for atoms in the sublattice A of graphene, originated because electrons can go from an A atom to the adatom right on top and then go back to the same atom A. This energy is weighted by the effective parameter t_{1a}. The third and fourth terms are both weighted by the parameters t_{1b} and t_{2b}. The third term is given by electrons that hop from B to the adatom and then go back to the same atom B, which results in an onsite term. On the other hand, and contributing to the same order, we have the fourth term that corresponds to electrons that go from B to the adatom and then to another second nearest neighbors atom B. We denote the annihilation(creation) operators for A and B sublattices by $a_i(a_i^\dagger)$ and $b_j(b_j^\dagger)$. $\tilde{\mathbf{n}}_{il}^B(\tilde{\mathbf{n}}_{il}^A)$ and $\tilde{\mathbf{n}}_{jl}^B$ are vectors that go from B(A) atom to the adatom. The subscript i is used for

Journal of Physics: Conference Series 2238 (2022) 012003 doi:10.1088/1742-6596/2238/1/012003

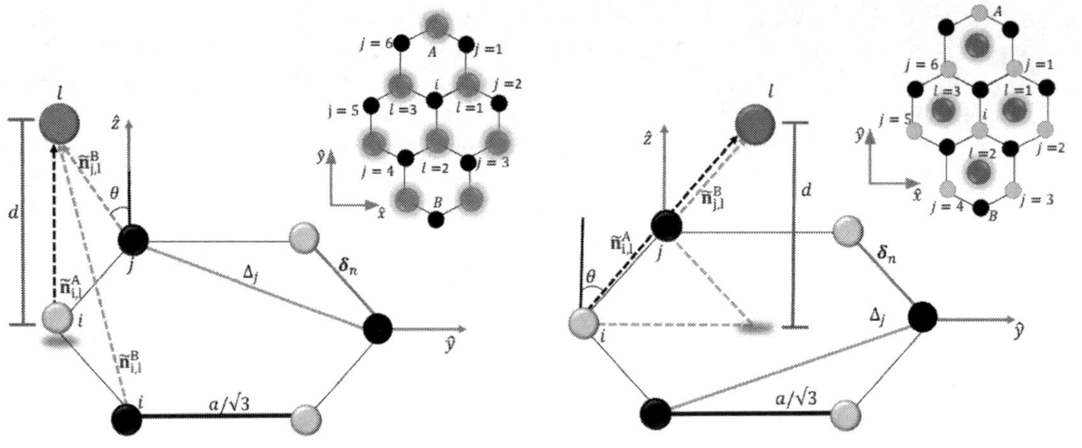

(a) Top configuration (b) Hollow configuration

Figure 1: (Color online)Schematic illustration of high symmetry adsorption sites for Alkali metal adatoms on graphene. a)Top configuration: the adatom (denoted by l) is placed on Top of the sublattice A of graphene at the vertical distance d. b) Hollow configuration: The l adatom is placed at the center of the hexagonal graphene rings at the distance d. In both illustrations graphene crystal structure shows the two interpenetrating lattices labeled as A (yellow-light) and B (black-dark) separated by the distance $a/\sqrt{3}$, being $\delta_n, n = 1; 2; 3$ the vector to 1st neighbors A(B)-B(A), and Δ_j, j=1,...,6 the A(B)-A(B) vector. $\widetilde{\mathbf{n}}_{il}$ and $\widetilde{\mathbf{n}}_{jl}$ are vectors that go from carbon atoms, A and B, to the adatoms l.

the reference atom B(A), the subscript j is used for its second neighbor in the graphene plane, and the subscript l is used for the alkali metal adatoms.

Equation (1) can be transformed to the reciprocal space as:

$$\begin{aligned} H_{\mathrm{T}}(\boldsymbol{k}) =& -(\gamma_0 + t_0)\sum_{\boldsymbol{k}}\left(f(\boldsymbol{k})A^\dagger_{\boldsymbol{k}}B_{\boldsymbol{k}} + f^*(\boldsymbol{k})B^\dagger_{\boldsymbol{k}}A_{\boldsymbol{k}}\right) - t_{1a}\sum_{\boldsymbol{k}}A^\dagger_{\boldsymbol{k}}A_{\boldsymbol{k}} \\ & - t_{1b}\sum_{\boldsymbol{k}}\left(1 + h(\boldsymbol{k})\right)B^\dagger_{\boldsymbol{k}}B_{\boldsymbol{k}} - t_{2b}\sum_{\boldsymbol{k}}\left(3b^2 - \frac{b^2}{2}h(\boldsymbol{k})\right)B^\dagger_{\boldsymbol{k}}B_{\boldsymbol{k}}, \end{aligned} \tag{2}$$

where $f(\boldsymbol{k}) = \mathrm{e}^{ik_ya/\sqrt{3}} + 2\mathrm{e}^{-ik_ya/2\sqrt{3}}\cos(k_xa/2)$ and $h(\boldsymbol{k}) = 4\cos(ak_x/2)\cos(a\sqrt{3}k_y/2) + 2\cos(ak_x)$. $A^\dagger_{\boldsymbol{k}}(B^\dagger_{\boldsymbol{k}})$ and $A_{\boldsymbol{k}}(B_{\boldsymbol{k}})$ are operators for creating and annihilating particles with quasi-momentum $\boldsymbol{k}$ defined for the A(B) sublattices.

2.2. Hollow Configuration

For alkali metals in the H configuration, the corresponding tight-binding Hamiltonian for graphene may be written in real space as:

$$\begin{aligned} H_{\mathrm{H}} =& \sum_{\langle ij\rangle}(a^\dagger_i b_j + b^\dagger_j a_i)(-\gamma_0 + h_0 + h_1(\widetilde{\boldsymbol{n}}^A_{il}\cdot\widetilde{\boldsymbol{n}}^B_{jl})) + \sum_{\langle\langle ij\rangle\rangle}(a^\dagger_i b_j + b^\dagger_j a_i)(h_0 + h_1(\widetilde{\boldsymbol{n}}^A_{il}\cdot\widetilde{\boldsymbol{n}}^B_{jl})) \\ & + \sum_i\left[a^\dagger_i a_i(h_0 + h_1(\widetilde{\boldsymbol{n}}^A_{il}\cdot\widetilde{\boldsymbol{n}}^A_{il})) + b^\dagger_i b_i(h_0 + h_1(\widetilde{\boldsymbol{n}}^B_{il}\cdot\widetilde{\boldsymbol{n}}^B_{il}))\right] + \sum_{\langle\langle ij\rangle\rangle}\left[(a^\dagger_i a_j + a^\dagger_j a_i)\right. \\ & \left.(h_0 + h_1(\widetilde{\boldsymbol{n}}^A_{il}\cdot\widetilde{\boldsymbol{n}}^A_{jl})) + (b^\dagger_i b_j + b^\dagger_j b_i)(h_0 + h_1(\widetilde{\boldsymbol{n}}^B_{il}\cdot\widetilde{\boldsymbol{n}}^B_{jl}))\right], \end{aligned} \tag{3}$$

where, as before, the first term is the kinetic contribution to nearest neighbors, corresponding to the pristine graphene hopping, perturbed by an additional term that includes the effect of hopping of electrons from the atom A(B) to the adatom and then to a nearest neighbor B(A). Contributing to the same order in the perturbation, in the second term we have also a kinetic term that corresponds to hops from A(B) to the adatom and then to a third neighbor B(A). The third term are onsite energies for A and B given by electronic transitions from A(B) to the adatom and then back to the same A(B) atom. Finally, the fourth term is given by electrons that go from A(B) to the adatom and then to a second neighbor A(B). All the energies in this Hamiltonian are written in terms of the effective parameters h_0 and h_1. The vectors $\tilde{\mathbf{n}}_{il}^{A}(\tilde{\mathbf{n}}_{il}^{B})$ and $\tilde{\mathbf{n}}_{jl}^{A}(\tilde{\mathbf{n}}_{jl}^{B})$ go from A(B) to the adatom as shown in Fig. 1b. The subscript i is used for the reference atom A(B), the subscript j is used for its second neighbor B, and the subscript l is used for the alkali metal atoms.

Equation (3) can be transformed to the Fourier space to obtain:

$$\begin{aligned} H_{\mathrm{H}}(\boldsymbol{k}) =& q_1 \sum_{\boldsymbol{k}} \left(f(\boldsymbol{k}) A_{\boldsymbol{k}}^{\dagger} B_{\boldsymbol{k}} + f^*(\boldsymbol{k}) B_{\boldsymbol{k}}^{\dagger} A_{\boldsymbol{k}} \right) + q_2 \sum_{\boldsymbol{k}} \left(g(\boldsymbol{k}) A_{\boldsymbol{k}}^{\dagger} B_{\boldsymbol{k}} + g^*(\boldsymbol{k}) B_{\boldsymbol{k}}^{\dagger} A_{\boldsymbol{k}} \right) \\ &+ p_1 \sum_{\boldsymbol{k}} (A_{\boldsymbol{k}}^{\dagger} A_{\boldsymbol{k}} + B_{\boldsymbol{k}}^{\dagger} B_{\boldsymbol{k}}) + p_2 \sum_{\boldsymbol{k}} (h(\boldsymbol{k}) + h^*(\boldsymbol{k}))(A_{\boldsymbol{k}}^{\dagger} A_{\boldsymbol{k}} + B_{\boldsymbol{k}}^{\dagger} B_{\boldsymbol{k}}), \end{aligned} \tag{4}$$

where $g(\boldsymbol{k}) = 2\mathrm{e}^{ik_y a/\sqrt{3}} \cos(k_x a) + \mathrm{e}^{-2ik_y a/\sqrt{3}}$. We define $q_1 \equiv -\gamma_0 + h_0 + h_1 b^2$, $q_2 \equiv h_0 - h_1 b^2$, $p_1 \equiv h_0 + 3b^2 h_1$, and $p_2 \equiv h_0 - \frac{b^2}{2} h_1$.

3. Band structure and the density of states

The band structure and density of states (DOS) for the T and H configurations can be computed using the Green's functions found by the equation of motion method [33, 34, 35]. With this end, there are defined the two Green's functions $\mathrm{G}_{\boldsymbol{k}} = -i\langle A_{\boldsymbol{k}} A_{\boldsymbol{k}}^{\dagger} \rangle$, and $\mathrm{F}_{\boldsymbol{k}} = -i\langle B_{\boldsymbol{k}} A_{\boldsymbol{k}}^{\dagger} \rangle$.

3.1. Top configuration

We apply Green's functions approach to calculate the equation of the system for alkali metals in the T configuration

$$\begin{aligned} s\mathrm{G}_{\boldsymbol{k}} + rf(\boldsymbol{k})\mathrm{F}_{\boldsymbol{k}} = 1; \\ \left(\omega + t_{1b} + t_{1b} h(\boldsymbol{k}) + 3t_{2b} b^2 - \frac{b^2}{2} t_{2b} h(\boldsymbol{k})\right) \mathrm{F}_{\boldsymbol{k}} + rf^*(\boldsymbol{k})\mathrm{G}_{\boldsymbol{k}} = 0, \end{aligned} \tag{5}$$

where $s \equiv \omega + t_{1a}$ and $r \equiv \gamma_0 + t_0$.

Solving the previous system of equations (5) we obtain the expressions for the Green's functions,

$$\begin{aligned} \mathrm{G}_{\boldsymbol{k}} &= \frac{-\left[2t_{1b}\left(1 + h(\boldsymbol{k})\right) + b^2 t_{2b}\left(-6 + h(\boldsymbol{k})\right) - 2\omega\right]}{\{\left[-2\left(1 + h(\boldsymbol{k})\right) t_{1b} + b^2\left(-6 + h(\boldsymbol{k})\right) t_{2b} - 2\omega\right] + 2f(\boldsymbol{k}) r^2 f^*(\boldsymbol{k})\}}; \\ \mathrm{F}_{\boldsymbol{k}} &= \frac{2rf(\boldsymbol{k})}{\{\left[-2\left(1 + h(\boldsymbol{k})\right) t_{1b} + b^2\left(-6 + h(\boldsymbol{k})\right) t_{2b} - 2\omega\right] + 2f(\boldsymbol{k}) r^2 f^*(\boldsymbol{k})\}}. \end{aligned} \tag{6}$$

3.2. Hollow configuration

For the H configuration we have the following system of equations:

$$\begin{aligned} \left(\omega - h_0 - 3h_1 b^2 - h(\boldsymbol{k})p\right) \mathrm{G}_{\boldsymbol{k}} - \left(qf(\boldsymbol{k}) + g(\boldsymbol{k})v\right) \mathrm{F}_{\boldsymbol{k}} = 1; \\ \left(\omega - h_0 - 3h_1 b^2 - h(\boldsymbol{k})p\right) \mathrm{F}_{\boldsymbol{k}} - \left(qf^*(\boldsymbol{k}) + g^*(\boldsymbol{k})v\right) \mathrm{G}_{\boldsymbol{k}} = 0, \end{aligned} \tag{7}$$

which can be solved in order to have the Green's functions,

$$
\begin{aligned}
\mathrm{G}_{\boldsymbol{k}} &= -\frac{(t+h(\boldsymbol{k})p-\omega)}{\left[(t+h(\boldsymbol{k})p-\omega)^2 - qf^*(\boldsymbol{k})\left(f(\boldsymbol{k})q+g(\boldsymbol{k})v\right) - vg^*(\boldsymbol{k})\left(f(\boldsymbol{k})q+g(\boldsymbol{k})v\right)\right]}; \\
\mathrm{F}_{\boldsymbol{k}} &= \frac{(qf^*(\boldsymbol{k})+vg^*(\boldsymbol{k}))}{\left[(t+h(\boldsymbol{k})p-\omega)^2 - qf^*(\boldsymbol{k})\left(f(\boldsymbol{k})q+g(\boldsymbol{k})v\right) - vg^*(\boldsymbol{k})\left(f(\boldsymbol{k})q+g(\boldsymbol{k})v\right)\right]}.
\end{aligned}
\tag{8}
$$

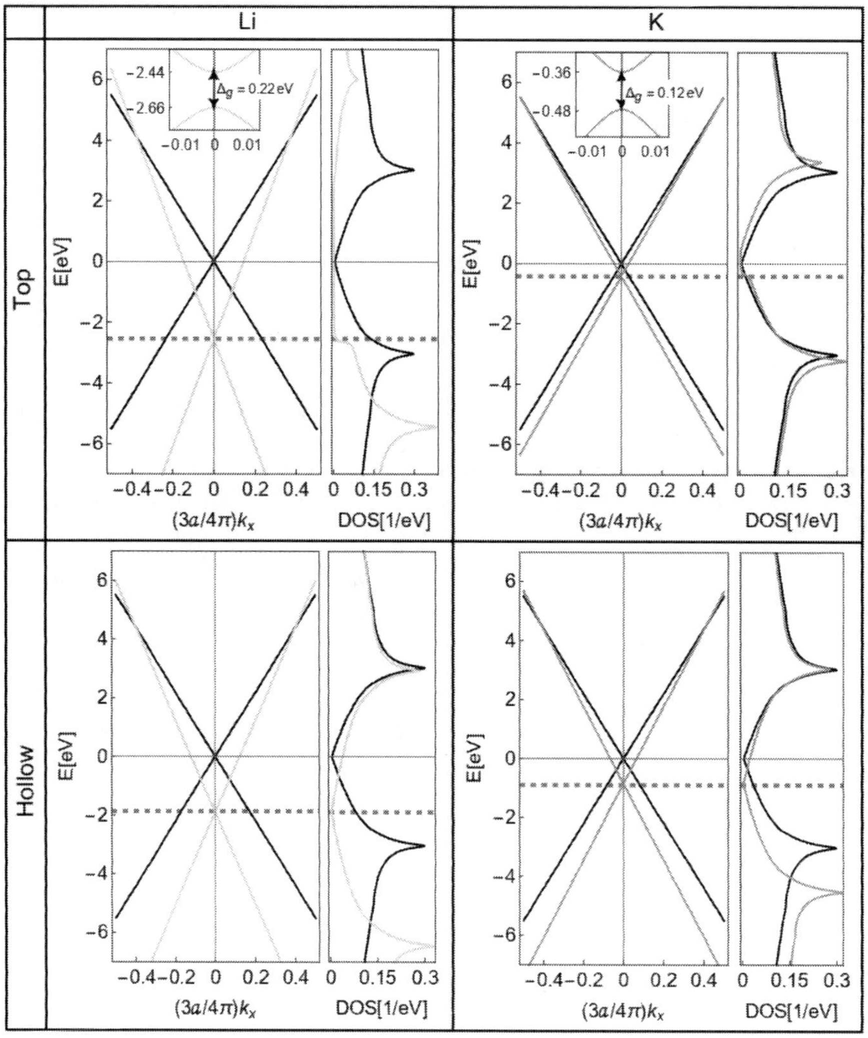

Figure 2: (Color online) Band Structure and corresponding density of states of graphene with adsorbed Li and K. **Top-left:** T configurations for Li adatom. **Top:** T configurations for K (left) and Li (right) adatoms. The inset above shows a band-gap of $\Delta_g = 0.22$ eV for Li and $\Delta_g = 0.12$ eV for K. The dashed gray line gives a shift of around -2.55 eV for Li and -0.42 eV for K. **Bottom:** H configuration for Li(left) and K(right) adatoms. In this case there is no gap between the valence and conduction bands, with a shifting of -1.89 eV for Li and -0.9 eV for K, represented by the gray-dashed line.

The band structures for the graphene π orbitals are calculated by finding the poles of the

Green's functions of Eqs. (6) and (8), which are expanded around the high symmetry $K_{\pm}$ points of the graphene first Brillouin zone. The corresponding densities of states are calculated by performing the trace of the spectral function as $\rho(\omega) = -\frac{1}{\pi}\Im G^r_{\boldsymbol{k}}$, with $G^r_{\boldsymbol{k}}$ as the well-known retarded Green function [33, 34, 35]. Fig. 2 shows the band structures and densities of states for both configurations, T and H, for Li and K on graphene. For the T case, Fig. 2(Top), we have a shifting of the Fermi level of −2.55 eV for Li(left) and −0.42 eV for K(right), represented by the horizontal gray(dashed) line, with a small gap of 0.22 eV and 0.12 eV for Li and K respectively, as can be seen in the insets in this figure. Accordingly, the densities of states show the same shifting of the Fermi level, with a small deformation in the minimum given by the gap. For the H case depicted in Fig. 2(Bottom), we have gapless dispersions with a shifting of −1.89 eV and −0.9 eV for Li and K respectively.

The symmetry of every configuration plays a significant role in the shape of the bands and densities of states. For the T configuration, we have that the adatom on top of the A sublattice breaks the sublattice symmetry, originating the gap, while for the H system, we have not this effect because the adatom is in the center of the hexagons, preserving the symmetry. On the other hand, the nature of the adatom is also important, we can see that K produces a smaller gap, a shorter shifting of the Fermi level, and a smaller change in the Fermi velocity compared to Li; this occurs because being a smaller atom, Li-graphene equilibrium distances are smaller compared to the K-graphene distances (see Table 1). Therefore, Li has a stronger hybridization with the p_z orbitals of graphene (see Table 2), having then a stronger influence in its electronic properties.

The effective parameters: t_0, t_{1a}, t_{1b}, t_{2b}, h_0, and h_1, used to calculate the band structures and densities of states shown in Fig. 2, are fitted in order to reproduce the bands given in the literature calculated by DFT [27], and their values are indicated in the Table 2.

Table 2: Parameters value of the Li and K on graphene for T and H configuration. The h_{0T}, h_{1a}, h_{1b}, and h_{2b} are parameter for the T configuration, and are coefficients given in terms of the site energies of the p orbitals of graphene. The h_0 and h_1 are parameter used for the H configuration.

Parameter(eV)	t_0	t_{1a}	t_{1b}	t_{2b}	h_0	h_1
Li	1.88	2.66	1.39	3.57	−0.1	−1.21
K	0.23	0.36	0.08	0.73	0.01	−0.91

4. Conductance

The numerical calculations for the conductance were performed by using the Kwant package[36] of Python. Kwant allows to build 2D crystal structures using lattice parameters. A nanorribon made of 170 × 340 graphene unit cells, each of them containing two atoms, has approximately 115, 600 carbon atoms. The adatoms contributions are included through the renormalized onsite and hopping energy values. This configuration gives an approximately 40 nm wide graphene nanoribbon with arm-chair edges. The model is based on tight binding Hamiltonians for: pristine graphene, and for Li and K on graphene in the T (Eq. 1) and H (Eq. 3) configurations considering the first and second nearest neighbors contribution. Then, two normal semi-infinite (defined width, infinite lenght) graphene leads or contacts were added to the ribbon and the conductance

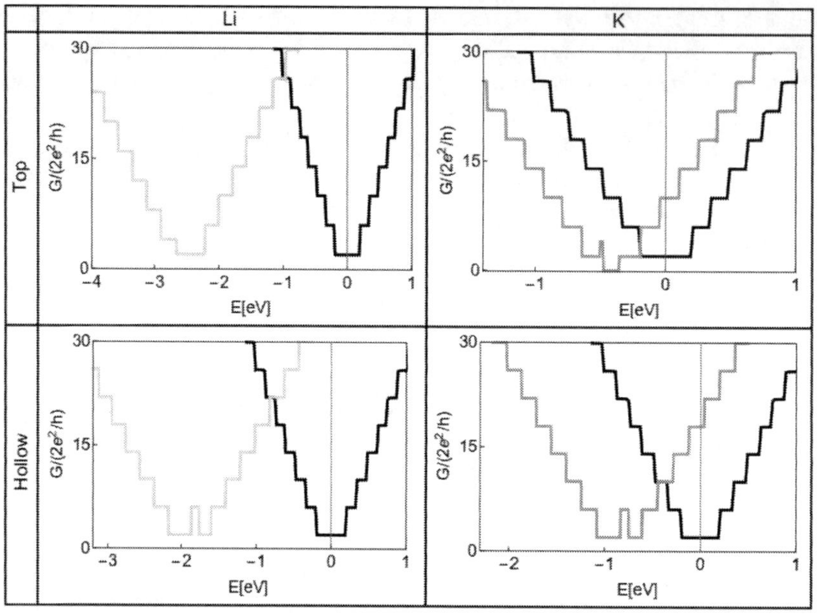

Figure 3: (Color online)Conductance of graphene with adsorbed Li and K. **Top-left:** T configurations for Li adatom. Top Li has shift of −2.55 eV with respect to the pristine. **Top-right:** T configurations for K adatom, with a shift of −0.42 eV. with respect to the pristine. **Bottom-left:** H configurations for Li with a shift of −1.89 eV. **Bottom-right:** H configurations for K adatom, with a shift of −0.9 eV with respect to pristine.

is calculated using *kwant.greens_function* method to obtain the transmission between the two leads as a function of the energy in eV.

The results of a quantified conductance for pristine graphene are in accordance to literature[37, 38]. As shown in Figure 3 the conductance is not degraded by the adatoms, preserving its symmetric behavior and the linear increasing with the absolute value of the energy, compared with the pristine graphene case. All the curves are shifted in the same energy values as the bands in Figure 2, as expected. Also, the perturbation of the Fermi velocities observed in the bands is translated to the V-shaped conductance, and the broadness of the lowest plateau increases for the graphene with adatoms as reported in the literature [39]. For the Top configurations the gap in the spectrum, should produce a suppression of the conductance around the charge neutrality point, effect that is observed for K but not for Li. We consider this discrepancy is related to size effects of the systems simulated and that further calculations are necessary to deepen their comprehension (see reference [40] where the appearance of dips and peaks in the conductance depending on the aspect ratio of the sample is discussed).

5. Summary and Conclusions

Starting from a TB Hamiltonian model for p_z orbitals of graphene perturbed by the interaction with Li and K in the configurations T and H, we have calculated the band structures and densities of states. The procedure followed avoids the diagonalization of the Hamiltonians, by finding the corresponding Green's functions for each configuration by the equation of motion method. Afterwards, a computational model based on Kwant which was built with the same Hamiltonians in the real space presented in section 2, was used to calculate the quantum conductance for a 40 nm wide graphene armchair nanoribbon.

 2238 (2022) 012003 doi:10.1088/1742-6596/2238/1/012003

Our results show that graphene π bands are down shifted by the interaction with Li and K for the T and H configurations. In all the cases considered the dispersions maintain almost the same Fermi velocity as pristine graphene, indicating that the graphene mobility and electronic properties are minimally degraded by the presence of the adatoms for the concentrations considered in this work. Another interesting characteristic is the opening of a gap for the T configuration, produced by the breaking of the AB symmetry that generates the adatoms on top of the A graphene sublattice. The conductance shows the same shifting in energy than the bands, and a growing with the absolute value of the energy, almost at the same rate as pristine graphene, which indicates that the quantum conductance is also preserved.

The analytical expressions for the Hamiltonians and their treatment for finding the bands and the DOS by the Green's functions equation of motion method, can be used in future works to make a more comprehensive study of the quantum properties, including the transmission through the ribbons, other kinds of boundaries, putting adatoms only in a central region of the ribbon, and even changing the concentration of the adatoms.

Acknowledgements
C.V. is thankful with the organizers of the Physics Meeting EPN 2021, held at the Escuela Politécnica Nacional for their excellent administrative and logistic work. M.P., C.V., and R.V. are grateful to the project PE2DMAT from Yachay Tech University for providing support for the realization of this work.

Appendix A. Reciprocal space Hamiltonians

Appendix A.1. Top Hamiltonian

Here we develop the conversion to the reciprocal space of the tight binding Hamiltonian for the Top configuration (equation (1)). $\widetilde{\boldsymbol{n}}_{il}^{A}$ is a vector in the plane that goes from the atom **A** of graphene to the adatom. In this configuration, $\widetilde{\boldsymbol{n}}_{il}^{A} = 0\hat{x} + 0\hat{y}$ (Fig. 1a), then, some terms will not contribute to the final result.

In the first term $a_i^\dagger$ and b_j are the annihilation and creation operators at the Ap_z and Bp_z orbitals of graphene:

$$\begin{aligned}
-(\gamma_0 + t_0)\sum_{\langle i,j\rangle} a_i^\dagger b_j &= -(\gamma_0 + t_0)\sum_{\boldsymbol{k},\boldsymbol{k}'}\frac{1}{N^2}\sum_{\langle i,j\rangle} \mathrm{e}^{-i\boldsymbol{k}\boldsymbol{x}_i} A_{\boldsymbol{k}}^\dagger B_{\boldsymbol{k}'} \mathrm{e}^{i\boldsymbol{k}'(\boldsymbol{x}_j)} \\
&= -(\gamma_0 + t_0)\sum_{\boldsymbol{k},\boldsymbol{k}'}\frac{1}{N^2}\sum_{i} \mathrm{e}^{i(\boldsymbol{k}'-\boldsymbol{k})(\boldsymbol{x}_i)}\left[\mathrm{e}^{i\boldsymbol{k}'\vec{\delta}_1} + \mathrm{e}^{i\boldsymbol{k}'\boldsymbol{\delta}_2} + \mathrm{e}^{i\boldsymbol{k}'\boldsymbol{\delta}_3}\right] A_{\boldsymbol{k}}^\dagger B_{\boldsymbol{k}'} \qquad (A.1)\\
&= -(\gamma_0 + t_0)\sum_{\vec{k}} f(\boldsymbol{k}) A_{\boldsymbol{k}}^\dagger B_{\boldsymbol{k}}.
\end{aligned}$$

The second term is,

$$\begin{aligned}
-t_{1a}\sum_i a_i^\dagger a_i &= -t_{1a}\sum_{\boldsymbol{k},\boldsymbol{k}'}\frac{1}{N^2}\sum_i \mathrm{e}^{i(\boldsymbol{k}'-\boldsymbol{k})\boldsymbol{x}_i} A_{\boldsymbol{k}}^\dagger A_{\boldsymbol{k}'} \\
&= -t_{1a}\sum_{\boldsymbol{k},\boldsymbol{k}'}\delta_{\boldsymbol{k},\boldsymbol{k}'} A_{\boldsymbol{k}}^\dagger A_{\boldsymbol{k}'} = -t_{1a}\sum_{\boldsymbol{k}} A_{\boldsymbol{k}}^\dagger A_{\boldsymbol{k}}.
\end{aligned} \qquad (A.2)$$

The third term corresponds to the atom **B**, it takes into account the hopping from **B** to the adatom, and vice versa. It also takes into account the hopping from **B** to the adatom and from the adatom to another atom **B** to second neighbors. We begin with the case $i = j$ and $\vec{\Delta}_j = \vec{0}$

$$
\begin{aligned}
-t_{2b}\sum_{l=1}^{3}\widetilde{\boldsymbol{n}}_{xil}^{B}\cdot\widetilde{\boldsymbol{n}}_{yil}^{B}\sum_{i}b_i^{\dagger}b_i =& - t_{2b}\Big[((\widetilde{\boldsymbol{n}}_{x1}^{B})^2+(\widetilde{\boldsymbol{n}}_{y1}^{B})^2)+((\widetilde{\boldsymbol{n}}_{x2}^{B})^2+(\widetilde{\boldsymbol{n}}_{y2}^{B})^2)+((\widetilde{\boldsymbol{n}}_{x3}^{B})^2 \\
&+(\widetilde{\boldsymbol{n}}_{y3}^{B})^2)\Big]b_i^{\dagger}b_i = -3t_{2b}b^2\sum_{\boldsymbol{k}}B_{\boldsymbol{k}}^{\dagger}B_{\boldsymbol{k}}.
\end{aligned}
\tag{A.3}
$$

The contributions to second neighbors for $i \neq j$ con $j = 1...6$; $\vec{\Delta}_j \neq 0$, is:

$$
\begin{aligned}
-t_{1b}\sum_{\langle\langle i,j\rangle\rangle}b_i^{\dagger}b_j =& - t_{1b}\sum_{\boldsymbol{k},\boldsymbol{k}'}\frac{1}{N^2}\sum_{\langle i,j\rangle}\mathrm{e}^{-i\boldsymbol{k}\boldsymbol{x}_i}B_{\boldsymbol{k}}^{\dagger}B_{\boldsymbol{k}'}\mathrm{e}^{i\boldsymbol{k}'\boldsymbol{x}_j} \\
=& - t_{1b}\sum_{\boldsymbol{k}}\Big[\mathrm{e}^{i\boldsymbol{k}\boldsymbol{\Delta}_1}+\mathrm{e}^{i\boldsymbol{k}\boldsymbol{\Delta}_2}+\mathrm{e}^{i\boldsymbol{k}\boldsymbol{\Delta}_3}+\mathrm{e}^{i\boldsymbol{k}\boldsymbol{\Delta}_4}+\mathrm{e}^{i\boldsymbol{k}\boldsymbol{\Delta}_5}+\mathrm{e}^{i\boldsymbol{k}\boldsymbol{\Delta}_6}\Big]B_{\boldsymbol{k}}^{\dagger}B_{\boldsymbol{k}} \\
=& - t_{1b}\sum_{\boldsymbol{k}}h(\boldsymbol{k})B_{\boldsymbol{k}}^{\dagger}B_{\boldsymbol{k}},
\end{aligned}
\tag{A.4}
$$

and,

$$
\begin{aligned}
-t_{2b}\sum_{\langle\langle i,j\rangle\rangle}(\widetilde{\boldsymbol{n}}_{il}^{B}\cdot\widetilde{\boldsymbol{n}}_{il}^{B})b_i^{\dagger}b_j =& - t_{2b}\sum_{j=1}^{6}\sum_{l=1}^{3}(\widetilde{\boldsymbol{n}}_{xil}^{B}\widetilde{\boldsymbol{n}}_{xjl}^{B}+\widetilde{\boldsymbol{n}}_{yil}^{B}\widetilde{\boldsymbol{n}}_{yjl}^{B})\frac{1}{N^2}\sum_{\boldsymbol{k},\boldsymbol{k}'}\mathrm{e}^{-i\boldsymbol{k}\boldsymbol{x}_i}B_{\boldsymbol{k}}^{\dagger}\mathrm{e}^{i\boldsymbol{k}'\boldsymbol{x}_j}B_{\boldsymbol{k}'} \\
=& t_{2b}\frac{b^2}{2}\sum_{\boldsymbol{k}}h(\boldsymbol{k})B_{\boldsymbol{k}}^{\dagger}B_{\boldsymbol{k}}.
\end{aligned}
\tag{A.5}
$$

In the same way, we can calculate the expression for atom **B** concerning the effective parameter t_{1b}. Finally, substituting all the transformations presented above in the Hamiltonian (1), we obtain equation (2).

Appendix A.2. Hollow Hamiltonian

For alkali metals in the H configuration, the corresponding tight-binding Hamiltonian for graphene in the real space is given by equation (3). The transformations of this Hamiltonian to the reciprocal space is as follows,

$$
\begin{aligned}
h_1\sum_{\langle ij\rangle}(\widetilde{\boldsymbol{n}}_{il}^{A}\cdot\widetilde{\boldsymbol{n}}_{jl}^{B})a_i^{\dagger}b_j =& h_1\sum_{j=1}^{3}(\widetilde{\boldsymbol{n}}_{il}^{A}\cdot\widetilde{\boldsymbol{n}}_{jl}^{B})\frac{1}{N}\sum_{\boldsymbol{k}}\mathrm{e}^{-ik\vec{x}_i}A_{\vec{k}}^{\dagger}\frac{1}{N}\sum_{\boldsymbol{k}'}\mathrm{e}^{i\boldsymbol{k}'\boldsymbol{x}_j}B_{\boldsymbol{k}'} \\
=& h_1\Big[\sum_{\vec{k}}\sum_{l=1}^{3}(\widetilde{\boldsymbol{n}}_{il}^{A}\cdot\widetilde{\boldsymbol{n}}_{1l}^{B})\mathrm{e}^{i\boldsymbol{k}\boldsymbol{\Delta}_1}A_{\boldsymbol{k}}^{\dagger}B_{\boldsymbol{k}}+\sum_{\boldsymbol{k}}\sum_{l=1}^{3}(\widetilde{\boldsymbol{n}}_{il}^{A}\cdot\widetilde{\boldsymbol{n}}_{2l}^{B})\mathrm{e}^{i\boldsymbol{k}\boldsymbol{\Delta}_2}A_{\boldsymbol{k}}^{\dagger}B_{\boldsymbol{k}}+ \\
&\sum_{\boldsymbol{k}}\sum_{l=1}^{3}(\widetilde{\boldsymbol{n}}_{il}^{A}\cdot\widetilde{\boldsymbol{n}}_{3l}^{B})\mathrm{e}^{i\boldsymbol{k}\boldsymbol{\Delta}_3}A_{\boldsymbol{k}}^{\dagger}B_{\boldsymbol{k}}\Big] = h_1b^2\sum_{\boldsymbol{k}}f(\boldsymbol{k})A_{\boldsymbol{k}}^{\dagger}B_{\boldsymbol{k}}.
\end{aligned}
\tag{A.6}
$$

Then, for the atom A

$$
\begin{aligned}
h_1\sum_{i}(\widetilde{\boldsymbol{n}}_{il}^{A}\cdot\widetilde{\boldsymbol{n}}_{il}^{A})a_i^{\dagger}a_i =& h_1\sum_{\boldsymbol{k}}\Big(\widetilde{\boldsymbol{n}}_{i1}^{B}\cdot\widetilde{\boldsymbol{n}}_{i1}^{A}+\widetilde{\boldsymbol{n}}_{i2}^{A}\cdot\widetilde{\boldsymbol{n}}_{i2}^{A}+\widetilde{\boldsymbol{n}}_{i3}^{A}\cdot\widetilde{\boldsymbol{n}}_{i3}^{A}\Big)A_{\boldsymbol{k}}^{\dagger}A_{\boldsymbol{k}'} \\
=& 3b^2h_1\sum_{\boldsymbol{k}}A_{\boldsymbol{k}}^{\dagger}A_{\boldsymbol{k}}.
\end{aligned}
\tag{A.7}
$$

The contributions to second neighbors for $i \neq j$ with $j = 1...3$; $\boldsymbol{\Delta}_j \neq 0$.

$$\begin{aligned} h_0 \sum_{\langle\langle i,j\rangle\rangle} a_i^\dagger b_j =& h_0 \sum_{\boldsymbol{k},\boldsymbol{k}'} \frac{1}{N^2} \sum_i \mathrm{e}^{i(\boldsymbol{k}'-\boldsymbol{k})(\boldsymbol{x}_i)} \left[\mathrm{e}^{i\boldsymbol{k}'\boldsymbol{\Delta}_1} + \mathrm{e}^{i\boldsymbol{k}'\boldsymbol{\Delta}_2} + \mathrm{e}^{i\boldsymbol{k}'\boldsymbol{\Delta}_3}\right] A_{\boldsymbol{k}}^\dagger B_{\boldsymbol{k}'} \\ =& h_0 \sum_{\boldsymbol{k}} g(\boldsymbol{k}) A_{\boldsymbol{k}}^\dagger B_{\boldsymbol{k}}. \end{aligned} \tag{A.8}$$

For h_1 to second neighbors,

$$\begin{aligned} h_1 \sum_{\langle\langle ij\rangle\rangle} (\widetilde{\boldsymbol{n}}_{il}^A \cdot \widetilde{\boldsymbol{n}}_{jl}^B) a_i^\dagger b_j =& h_1 \sum_{\vec{k}} \sum_{j=1}^{3} (\widetilde{\boldsymbol{n}}_{il}^A \cdot \widetilde{\boldsymbol{n}}_{jl}^B) \mathrm{e}^{i\vec{k}\boldsymbol{\Delta}_j} A_{\boldsymbol{k}}^\dagger B_{\boldsymbol{k}} \\ =& h_1 \sum_{\boldsymbol{k}} \left[\left(\widetilde{\boldsymbol{n}}_{i1}^A \cdot \widetilde{\boldsymbol{n}}_{11}^B\right) \mathrm{e}^{i\boldsymbol{k}\boldsymbol{\Delta}_1} + \left(\widetilde{\boldsymbol{n}}_{i2}^A \cdot \widetilde{\boldsymbol{n}}_{22}^B\right) \mathrm{e}^{i\boldsymbol{k}\boldsymbol{\Delta}_2} + \right. \\ & \left. \left(\widetilde{\boldsymbol{n}}_{i3}^A \cdot \widetilde{\boldsymbol{n}}_{33}^B\right) \mathrm{e}^{i\boldsymbol{k}\boldsymbol{\Delta}_3} \right] A_{\boldsymbol{k}}^\dagger B_{\boldsymbol{k}} = -h_1 b^2 \sum_{\boldsymbol{k}} g(\boldsymbol{k}) A_{\boldsymbol{k}}^\dagger B_{\boldsymbol{k}}. \end{aligned} \tag{A.9}$$

The contributions for the second neighbors for $i \neq j$ con $j = 1...6$; $\boldsymbol{\Delta}_j \neq 0$.

$$\begin{aligned} h_0 \sum_{\langle\langle i,j\rangle\rangle} a_i^\dagger a_j =& h_0 \sum_{\boldsymbol{k},\boldsymbol{k}'} \frac{1}{N^2} \sum_{\langle i,j\rangle} \mathrm{e}^{-i\boldsymbol{k}\boldsymbol{x}_i} A_{\boldsymbol{k}}^\dagger A_{\boldsymbol{k}'} \mathrm{e}^{i\boldsymbol{k}'\boldsymbol{x}_j} \\ =& h_0 \sum_{\boldsymbol{k}} \left[\mathrm{e}^{ia/2(k_x+\sqrt{3}k_y)} + \mathrm{e}^{iak_x} + \mathrm{e}^{ia/2(k_x-\sqrt{3}k_y} + \right. \\ & \left. \mathrm{e}^{-ia/2(k_x+\sqrt{3}k_y} + \mathrm{e}^{-iak_x} + \mathrm{e}^{ia/2(-k_x+\sqrt{3}k_y} \right] A_{\boldsymbol{k}}^\dagger A_{\boldsymbol{k}} \\ =& h_0 \sum_{\boldsymbol{k}} h(\boldsymbol{k}) A_{\boldsymbol{k}}^\dagger A_{\boldsymbol{k}}. \end{aligned} \tag{A.10}$$

Finally,

$$\begin{aligned} h_1 \sum_{\langle\langle i,j\rangle\rangle} (\widetilde{\boldsymbol{n}}_{il}^A \cdot \widetilde{\boldsymbol{n}}_{jl}^A) a_i^\dagger a_j =& h_1 \sum_{j=1}^{6} (\widetilde{\boldsymbol{n}}_{xil}^A \widetilde{\boldsymbol{n}}_{xjl}^A + \widetilde{\boldsymbol{n}}_{yil}^A \widetilde{\boldsymbol{n}}_{yjl}^A) \frac{1}{N} \sum_{\boldsymbol{k}} \mathrm{e}^{-i\boldsymbol{k}\boldsymbol{x}_i} A_{\boldsymbol{k}}^\dagger \frac{1}{N} \sum_{\boldsymbol{k}'} \mathrm{e}^{i\boldsymbol{k}'\boldsymbol{x}_j} A_{\boldsymbol{k}'} \\ =& -h_1 \frac{b^2}{2} \sum_{\boldsymbol{k}} h(\boldsymbol{k}) A_{\boldsymbol{k}}^\dagger A_{\boldsymbol{k}}. \end{aligned} \tag{A.11}$$

In the same way, we can calculate the transformations for the atom B. Then, the reciprocal space Hamiltonian for the Hollow configuration is given by the equation (4).

Appendix B. Calculation of the Green Functions

In this section, we work with the expressions for Green's functions for $\mathrm{G}_{\boldsymbol{k}}$, and $F_{\boldsymbol{k}}$.

Appendix B.1. Top configuration

First we calculate $\mathrm{G}_{\boldsymbol{k}}$ using the equation of motion, $\omega \mathrm{G}_{\boldsymbol{k}} = \delta_{\boldsymbol{k}\boldsymbol{k}} + \langle [A_{\boldsymbol{k}}, H_T]; A_{\boldsymbol{k}}^\dagger \rangle$.
The commutator in this expression is:

$$[A_{\boldsymbol{k}}, H_T] = (\gamma_0 + t_0) f(\boldsymbol{k}) B_{\boldsymbol{k}} - h_{1a} A_{\boldsymbol{k}}. \tag{B.1}$$

The equation of motion for $\mathrm{G}_{\boldsymbol{k}}$ reduces to

$$(\omega + h_{1a})\mathrm{G}_{\boldsymbol{k}} = -(\gamma_0 + t_0)f(\boldsymbol{k})\langle B_{\boldsymbol{k}} A^{\dagger}_{\boldsymbol{k}}\rangle + 1. \tag{B.2}$$

In the same way, we work with the equation of motion $\omega\mathrm{F}_{\boldsymbol{k}} = \langle [B_{\boldsymbol{k}}, H_T]; A^{\dagger}_{\boldsymbol{k}}\rangle$. Calculating the commutator:

$$[B_{\boldsymbol{k}}, H_T] = -(\gamma_0 + t_0)f^*(\boldsymbol{k})A_{\boldsymbol{k}} - h_{1b}B_{\boldsymbol{k}}(1 + h(\boldsymbol{k})) - b^2 B_{\boldsymbol{k}}(3h_{2b} - \frac{b^2}{2}h(\boldsymbol{k})). \tag{B.3}$$

The final expression for the equation of motion $\mathrm{F}_{\boldsymbol{k}}$ is:

$$\Big(\omega + h_{1b}h_{1b}h(\boldsymbol{k}) + 3h_{2b}b^2 - h_{2b}\frac{b^2}{2}h(\boldsymbol{k})\Big)\mathrm{F}_{\boldsymbol{k}} = -(\gamma_0 + t_0)f^*(\boldsymbol{k})\mathrm{G}_{\boldsymbol{k}}. \tag{B.4}$$

Solving the previous system of equations, we find the expressions for the Green's functions, equation (6).

Appendix B.2. Hollow configuration
As explained in the latest section, we work with the equation of motion for $\omega\mathrm{G}_{\boldsymbol{k}} = \delta_{\boldsymbol{k}\boldsymbol{k}} + \langle [A_{\boldsymbol{k}}, H_H]; A^{\dagger}_{\boldsymbol{k}}\rangle$. The result of the commutator using the H_H is:

$$[A_{\boldsymbol{k}}, H_H] = q_1 f(\boldsymbol{k})B_{\boldsymbol{k}} + q_2 g(\boldsymbol{k})B_{\boldsymbol{k}} + p_1 A_{\boldsymbol{k}} + p_2(h(\boldsymbol{k}) + h^*(\boldsymbol{k}))A_{\boldsymbol{k}}. \tag{B.5}$$

The equation of motion for $\mathrm{G}_{\boldsymbol{k}}$ reduces to

$$\Big(\omega - p_1 - p_2 h(\boldsymbol{k})\Big)\mathrm{G}_{\boldsymbol{k}} = \Big(q_1 f(\boldsymbol{k}) + q_2 g(\boldsymbol{k})\Big)\langle B_{\boldsymbol{k}} A^{\dagger}_{\boldsymbol{k}}\rangle + 1. \tag{B.6}$$

In the same way, we work with the equation of motion for $\omega\mathrm{F}_{\boldsymbol{k}} = \langle [B_{\boldsymbol{k}}, H_T]; A^{\dagger}_{\boldsymbol{k}}\rangle$. Calculating the commutator we have:

$$[B_{\boldsymbol{k}}, H_H] = q_1 f^*(\boldsymbol{k})A_{\boldsymbol{k}} + q_2 g^*(\boldsymbol{k})A_{\boldsymbol{k}} + p_1 B_{\boldsymbol{k}} + p_2(h(\boldsymbol{k}) + h^*(\boldsymbol{k}))B_{\boldsymbol{k}}. \tag{B.7}$$

Having the final expression for the equation of motion $\mathrm{F}_{\boldsymbol{k}}$ as:

$$\Big(\omega - p_1 - p_2(h(\boldsymbol{k}) + h^*(\boldsymbol{k}))\Big)\mathrm{F}_{\boldsymbol{k}} = \Big(q_1 f^*(\boldsymbol{k}) + q_2 g^*(\boldsymbol{k})\Big)\mathrm{G}_{\boldsymbol{k}}. \tag{B.8}$$

Finally, the system of equations for the H_H is shown in equation (7). Solving this system we find the expressions for the Green's functions, equation (8).

References

[1] K. T. Chan, J. Neaton, M. L. Cohen, First-principles study of metal adatom adsorption on graphene, Physical Review B 77 (23) (2008) 235430.

[2] M. Peralta, E. Medina, F. Mireles, Induced exchange and spin-orbit effects by proximity in graphene on ni and co, arXiv preprint arXiv:1903.06959 (2019).

[3] X. Liu, C. Wang, Y. Yao, W. Lu, M. Hupalo, M. Tringides, K. Ho, Bonding and charge transfer by metal adatom adsorption on graphene, Physical Review B 83 (23) (2011) 235411.

[4] K. Nakada, A. Ishii, Migration of adatom adsorption on graphene using dft calculation, Solid State Communications 151 (1) (2011) 13–16.

[5] I. Žutić, A. Matos-Abiague, B. Scharf, H. Dery, K. Belashchenko, Proximitized materials, Materials Today 22 (2019) 85–107.

[6] Y. Mao, J. Yuan, J. Zhong, Density functional calculation of transition metal adatom adsorption on graphene, Journal of Physics: Condensed Matter 20 (11) (2008) 115209.

[7] Y. Zhou, X. T. Zu, F. Gao, H. Lv, H. Xiao, Adsorption-induced magnetic properties and metallic behavior of graphene, Applied Physics Letters 95 (12) (2009) 123119.
[8] A. Shick, A. Y. Denisov, Magnetism of 4f-atoms adsorbed on metal and graphene substrates, Journal of Magnetism and Magnetic Materials 475 (2019) 211–215.
[9] M. Peralta, L. Colmenarez, A. López, B. Berche, E. Medina, Ferromagnetic order induced on graphene by ni/co proximity effects, Physical Review B 94 (23) (2016) 235407.
[10] D. Marchenko, A. Varykhalov, M. Scholz, G. Bihlmayer, E. Rashba, A. Rybkin, A. Shikin, O. Rader, Giant rashba splitting in graphene due to hybridization with gold, Nature communications 3 (1) (2012) 1–6.
[11] A. G. Rybkin, A. A. Rybkina, M. M. Otrokov, O. Y. Vilkov, I. I. Klimovskikh, A. E. Petukhov, M. V. Filianina, V. Y. Voroshnin, I. P. Rusinov, A. Ernst, et al., Magneto-spin–orbit graphene: interplay between exchange and spin–orbit couplings, Nano letters 18 (3) (2018) 1564–1574.
[12] A. López, L. Colmenárez, M. Peralta, F. Mireles, E. Medina, Proximity-induced spin-orbit effects in graphene on au, Physical Review B 99 (8) (2019) 085411.
[13] H. Yildirim, A. Kinaci, Z.-J. Zhao, M. K. Chan, J. P. Greeley, First-principles analysis of defect-mediated li adsorption on graphene, ACS applied materials & interfaces 6 (23) (2014) 21141–21150.
[14] N. Kheirabadi, A. Shafiekhani, Graphene/li-ion battery, Journal of Applied Physics 112 (12) (2012) 124323.
[15] G. Profeta, M. Calandra, F. Mauri, Phonon-mediated superconductivity in graphene by lithium deposition, Nature Physics 8 (2012) 131–134.
[16] B. M. Ludbrook, G. Levya, P. Niggea, M. Zonnoa, M. Schneider, D. J. Dvorak, C. N. Veenstra, S. Zhdanovich, D. Wong, P. Dosanjh, C. Straßer, A. Stöhr, S. Forti, C. R. Ast, U. Starke, , A. Damascelli, Evidence for superconductivity in li-decorated monolayer graphene, PNAS 112 No 38 (2015) 11795–11799.
[17] A. Lugo-Solis, I. Vasiliev, Ab initio study of k adsorption on graphene and carbon nanotubes: Role of long-range ionic forces, Physical Review B 76 (23) (2007) 235431.
[18] S. Sangavi, N. Santhanamoorthi, S. Vijayakumar, Density functional theory study on the adsorption of alkali metal ions with pristine and defected graphene sheet, Molecular Physics 117 (4) (2019) 462–473.
[19] B. Oli, C. Bhattarai, B. Nepal, N. Adhikari, First-principles study of adsorption of alkali metals (li, na, k) on graphene, in: Advanced Nanomaterials and Nanotechnology, Springer, 2013, pp. 515–529.
[20] E. Olsson, G. Chai, M. Dove, Q. Cai, Adsorption and migration of alkali metals (li, na, and k) on pristine and defective graphene surfaces, Nanoscale 11 (12) (2019) 5274–5284.
[21] Y. Zhou, W. Chu, F. Jing, J. Zheng, W. Sun, Y. Xue, Enhanced hydrogen storage on li-doped defective graphene with b substitution: A dft study, Applied Surface Science 410 (2017) 166–176.
[22] N. Zheng, S. Yang, H. Xu, Z. Lan, Z. Wang, H. Gu, A dft study of the enhanced hydrogen storage performance of the li-decorated graphene nanoribbons, Vacuum 171 (2020) 109011.
[23] Y.-T. Lin, S.-Y. Lin, Y.-H. Chiu, M.-F. Lin, Alkali-created rich properties in grapheme nanoribbons: Chemical bondings, Scientific Reports 7 (2017) 1722.
[24] B. Peles-Lemli, D. Kánnár, J.-C. Nie, H. Li, S. Kunsagi-Mate, Some unexpected behaviour of the adsorption of alkali metal ions onto the graphene surface under the effect of external electric field, Journal of Chemical Physics 117, 41 (2013) 21509–21515.
[25] K.-H. Jin, S.-M. Choi, S.-H. Jhi, Crossover in the adsorption properties of alkali metals on graphene, Physical Review B 82 (2010) 033414.
[26] A. Lugo-Solis, I. Vasiliev, Ab initio study of k adsorption on graphene and carbon nanotubes: Role of long-range ionic forces, Physical Review B 76 (2007) 235431.
[27] C. S. Praveen, S. Piccinin, S. Fabris, Adsorption of alkali adatoms on graphene supported by the au/ni(111) surface, Physical Review B 92 (2015) 075403.
[28] S. Woo, S. Hemmatiyan, T. D. Morrison, K. D. D. Rathnayaka, I. F. Lyuksyutov, D. G. Naugle, Temperature-dependent transport properties of graphene decorated by alkali metal adatoms (li, k), Applied Physics Letters 111 (2017) 263502.
[29] J. Yang, Y. Yuan, G. Chen, First–principles study of potassium adsorption and diffusion on graphene, Molecular Physics 118 (2020) e1581291.
[30] S. P. Repetsky, I. G. Vyshyvana, E. Y. Kuznetsova, S. P. Kruchinin, Energy spectrum of graphene with adsorbed potassium atoms, International Journal of Modern Physics B 32 (2018) 1840030.
[31] T. M. Radchenko, V. A. Tatarenko, I. Y. Sagalianov, Y. I. Prylutskyy, P. Szroeder, S. Biniak, On adatomic-configuration-mediated correlation between electrotransport and electrochemical properties of graphene, Carbon 101 (2016) 37–48.
[32] M. Farjam, H. Rafii-Tabar, Energy gap opening in submonolayer lithium on graphene: Local density functional and tight-binding calculations, Physical Review B 79 (2009) 045417.
[33] M. M. Odashima, B. G. Prado, E. Vernek, Pedagogical introduction to equilibrium green's functions: condensed-matter examples with numerical implementations, Revista Brasileira de Ensino de Física 39 (1) (2017).

[34] R. M. Martin, L. Reining, D. M. Ceperley, Interacting electrons, Cambridge University Press, 2016.

[35] E. Economou, Green's Functions in Quantum Physics, Springer series in solid-state sciences, Yang Kuo-fan, 1979.
URL https://books.google.es/books?id=uI4fAQAAMAAJ

[36] C. W. Groth, M. Wimmer, A. R. Akhmerov, X. Waintal, Kwant: a software package for quantum transport, New Journal of Physics 16 (6) (2014) 063065.

[37] R. N. Sajjad, F. Tseng, K. M. M. Habib, A. W. Ghosh, Quantum transport at the dirac point: Mapping out the minimum conductivity from pristine to disordered graphene, Phys. Rev. B 92 (2015) 205408. doi:10.1103/PhysRevB.92.205408.
URL https://link.aps.org/doi/10.1103/PhysRevB.92.205408

[38] L. Parisi, G. G. Angilella, I. Deretzis, R. Pucci, A. La Magna, Role of h distribution on coherent quantum transport of electrons in hydrogenated graphene, Condensed Matter 2 (4) (2017) 37.

[39] C. Weeks, J. Hu, J. Alicea, M. Franz, R. Wu, Engineering a robust quantum spin hall state in graphene via adatom deposition, Physical Review X 1 (021001) (2011).

[40] J. R. Williams, D. A. Abanin, L. DiCarlo, L. S. Levitov, C. M. Marcus, Quantum hall conductance of two-terminal graphene devices, Physical Review B 80 (045408) (2009).

Structural and thermal properties of InSb:Mn with MnSb clusters

C A Durante Rincón[1], J González-Laprea[2], M E Mendoza Oliveros[3], J A Castro[1], E Quintana[1] and J R Fermín[1, 4]

[1] Laboratorio de Ciencia de Materiales, Departamento de Física, Facultad Experimental de Ciencias, Universidad del Zulia, Maracaibo, Venezuela
[2] Facultad de Ciencias Exactas y Naturales, Escuela de Ciencias Físicas y Matemática, Pontificia Universidad Católica del Ecuador, Quito, Ecuador, 1076
[3] Universidad Pedagógica y Tecnológica de Colombia. Avenida Central del Norte 39-115, Tunja, Colombia
[4] Escuela de Ingeniería Industrial, Facultad de Ingeniería, Universidad Rafael Urdaneta, Maracaibo, Venezuela

jegonzalezl@puce.edu.ec, durincarlos@gmail.com

Abstract. III-V family compounds doped with transition metals are promising materials for spintronic applications. Synthesis of an $In_{0.9}Mn_{0.1}Sb$ ingot was thus carried out by direct fusion of the stoichiometric mixture of the constituent elements, followed by controlled cooling. The ingot obtained showed *p*-type conductivity. Scanning Electron Microscopy (SEM) images show MnSb clusters in an InSb matrix doped with Mn, a result like that found when the compound is obtained using other techniques. Energy-dispersive X-ray spectroscopy (EDS) shows that the atomic ratio of the clusters is Mn/Sb = 0.896 ± 0.025, while the atomic ratio of the matrix is In/Sb = 1.013 ± 0.005. The indexation of the powder X-ray diffraction pattern at room temperature yielded a majority cubic phase of InSb doped with Mn, with a lattice parameter $a = 6.474173$ Å and cell unit volume $V = 271.36$ Å^3, while non-indexed reflections are associated with the presence of MnSb rich in Sb. The phase transition temperatures were obtained from differential thermal analysis (DTA) measurements on powder samples in evacuated quartz capsules. It can be observed that fusion of the InSb matrix doped with Mn occurs between 485 °C and 528 °C, unlike the congruent fusion of the InSb at 527.7 °C; while the fusion of the Sb-rich MnSb clusters occurs between 494 °C and 509 °C. These temperatures are lower than those reported for the Sb-rich side of the phase diagram of the Mn-Sb binary system, which shows a decrease in the thermal stability of the compounds. The estimated fusion enthalpies for InSb:Mn and antimony-rich MnSb are, respectively, 4.8 Kcal/mol and 117.4 Kcal/mol.

1. Introduction

Microelectronic devices and circuits use the charge property of the electron, while the *spin* property of an electron is used in data storage media, such as computer hard drives, Magneto Random Access Memory (MRAM), and other applications. The study of materials that will permit the exploitation of electron charge and spin for possible applications will allow for the development of so-called *spintronic devices* [1]. One requirement in making these devices is the availability of semiconductor materials that behave ferromagnetically at room temperature [2].

This has not been previously done because the semiconductors used, such as silicon and gallium arsenide (the bases of current technology), are non-magnetic. On the other hand, the magnetic fields that would have to be used to take advantage of spin directions are too high for daily use. While there are semiconductors, such as europium chalcogenide and spinels (ferro and ferrimagnetic), that have ferromagnetic properties, from the technological point of view they have little use: crystal growth is very difficult [3], and compatibility with Si and GaAs is low. As such, they are not ideal for spintronic applications [1].

One of the approaches to building a ferromagnetic semiconductor is the introduction of magnetic ions such as Mn, Cr, Co and Fe into non-magnetic semiconductors. In these ferromagnetic semiconductors, through substitution, part of the lattice is composed of magnetic atoms. These are therefore called diluted magnetic semiconductors (DMS) [1].

Ferromagnetism in III-V compounds has gained interest for its possible technological applications. With this focus, intensive research on III-V semiconductors doped with transition metals is underway [4].

III-V compounds doped with Mn are the most promising materials for spintronic applications, due to the fact they show ferromagnetism at a relatively high Curie Temperature (C_T = 173 K).

III-V antimony-based semiconductor materials offer great freedom and flexibility for the engineering and structural design of materials for high performance microelectronics, optoelectronic devices and integrated circuits. The various applications could include, among others, satellite communications, ultra-high speed/low power consumption integrated circuits, portable/mobile devices, environmental monitoring, chemical detection, biomedical diagnostics, and drug analysis [5].

InSb is a III-V semiconductor compound with direct energy gap whose properties were first published by H. Welker in 1952 [6]. It has the lowest melting point, the smallest energy gap and the highest electron mobility among III-V compounds. When the group III element is replaced by Mn in this compound, it provides local momentum, as well as a free hole for conduction. Holes concentration must therefore be increased in order to achieve a higher ferromagnetic order and a higher Curie Temperature C_T. This is done by doping the host semiconductor with a large amount of manganese. However, a greater amount of manganese increases the Curie Temperature, while also introducing secondary phases. In fact, although the *ab initio* calculation has been used to obtain the $In_{1-x}Mn_xSb$ system properties in the entire range of compositions [7], experimentally it has been noted that manganese solubility in InSb is limited. InSb and MnSb thus form an eutectic [8], and the presence of clusters with manganese has been found in solid polycrystalline $In_{1-x}Mn_xSb$ solutions obtained by fusion of InSb with Mn [9, 10] and by fusion of the constituent elements with argon arc [11].

On the one hand, obtaining $In_{1-x}Mn_xSb$ system alloys with high Mn concentrations has not been possible using techniques such as the fusion of InSb with Mn [9, 10] or the fusion of the constituent elements with the argon arc [11]. On the other hand, the usefulness of a material in spintronic applications depends on its magnetic properties, which are closely related to its morphological, structural and thermal properties. With this in mind, in this work the method of direct fusion of the constituent elements and subsequent controlled cooling is for the first time used to obtain the compound in question; and its morphological, structural and thermal characterization is carried as an initial contribution towards understanding future magnetic measurements and their possible application in spintronics. The composition x = 0.1 was selected as an initial test, sufficiently far from doping concentrations. The morphological, structural and thermal characterization of the product obtained was carried out using Scanning Electron Microscopy (SEM), Energy-dispersive X-ray spectroscopy (EDS), X-ray Powder Diffraction and Differential Thermal Analysis (DTA). Morphology and composition, lattice parameters, phase transition temperatures and enthalpies of fusion for the obtained material are reported.

2. Experimental Details

Ingot synthesis was performed in a semiconductor quality quartz tube capsule with a 14.2 mm external diameter, a 12.0 mm internal diameter and a 2.2 mm wall. The capsule was made cylindrical, with a length of 15 cm and a conical bottom, thus guaranteeing a great enough volume to withstand high vapor pressures. It underwent a cleaning and thermal treatment process prior to use in the synthesis. Its interior was cleaned with a brush, quartz detergent and water, before rinsing with deionized water. Afterwards it was washed with acetone to eliminate possible organic residues, rinsed again with deionized water, and dried for 10 min in a stove at 120 °C. Finally, it was filled with an acid solution (5% HF, 33% HNO_3) for 10 minutes to eliminate inorganic residues and rinsed with deionized water before being again dried in an oven for 10 min at 120 °C.

Adhesion of the ingot to the quartz capsule wall, caused by the presence of O_2 or H_2O, produces cracks in the ingot, limits crystal size, can cause the capsule to fracture at high temperatures and makes it difficult for the ingot to be removed in one piece. To avoid this, the inner wall of the capsule was coated with graphite using acetone pyrolysis.

To obtain the ingot, indium (In) pieces with a purity degree of 99.999%, antimony (Sb) grains with a purity degree of 99.9999%, and manganese (Mn) powder with a purity degree of 99.99% were used. The necessary quantities of each element were calculated to obtain a 5 g ingot. The elements were then weighed using an analytical balance with a precision to 0.0001 g. The elements were added to the graphited capsule and it was placed in a high vacuum system (~10^{-6} torr) for 24 hours (see Figure 1), then sealed.

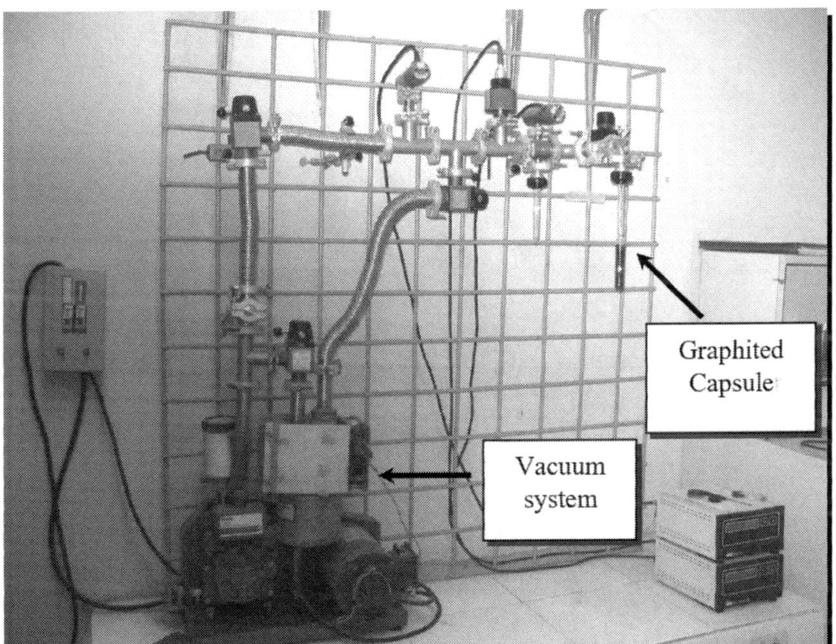

Figure 1. Synthesis capsule subjected to the vacuum process for subsequent sealing.

To elaborate the synthesis program shown in Figure 2, fusion points of the constituent elements and phase transitions temperature must be taken into account in the controlled cooling process. The room temperature is used as starting point and heating and cooling rates in °C/h are chosen to allow for adequate mixing of the molten materials, as well as a slow step for the solidification temperature to avoid undesired growth phases. Once growth is completed, the ingot is subjected to a thermal annealing process to improve its quality. A single zone vertical furnace with a chromel/chromel-alumel thermocouple controlled by an Omega CN 3000 controller/programmer was used.

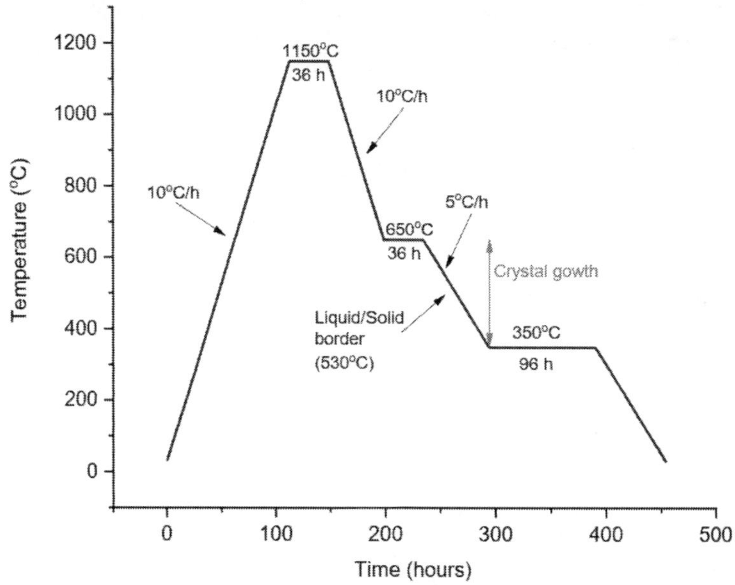

Figure 2. $In_{0.9}Mn_{0.1}Sb$ ingot growth program.

The ingot obtained is shown in Figure 3. Its characteristics are shown in Table 1. The simple thermal probe test showed *p*-type conductivity.

Table 1. Ingot Characteristics.

Height	Maximum Diameter	Color	Shape	Weight	Consistency	Conductivity
15.20 ± 0.01 mm	11.40 ± 0.01 mm	Silver	Conical	4.5736 ± 0.0001 g	Solid, no cracks, homogeneous	*p*-type

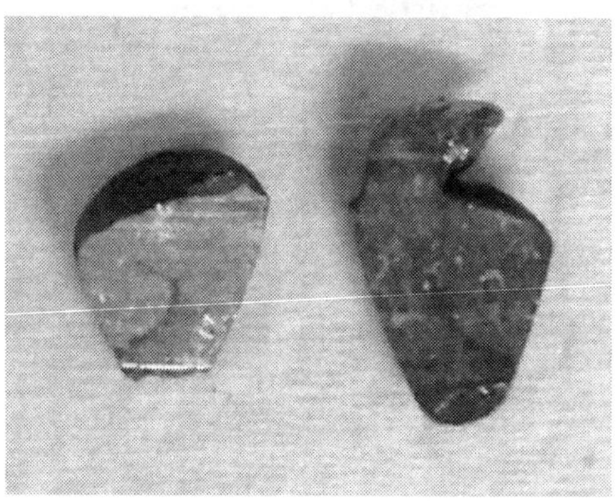

Figure 3. $Mn_{0.9}In_{0.1}Sb$ ingot

A FEI-Inspect S thermionic emission scanning electron microscope with a tungsten filament operated at 15 KV under high vacuum conditions was used to obtain compound photomicrographs. Backscattered electron detector (BSED) images were obtained to visualize contrast by atomic number

(Z). The composition of different regions was determined by Energy-dispersive X-ray spectroscopy using an EDAX solid-state detector. Results were analyzed using the Genesis Spectrum program, version 5.21.

X-ray powder diffraction at room temperature was carried out on a PANalytical X'Pert PRO diffractometer with copper anode ($\lambda = 1.540562$ Å) and Bragg-Brentano geometry. Measurement was carried out in a 2θ interval between 5° and 100° with a step size of 0.0170°, while the measurement step time was 1.9050 s/step.

The DTA measurements at different heating/cooling rates were obtained with a model DTA-50 Shimadzu device, equipped with Pt-Pt/10%Rh thermocouples and controlled by a computer running TA-60WS Software. The sample (82.4 mg) was ground in an agate mortar and crushed to powder. The crushed sample was placed in a quartz capsule that was vacuum sealed (~10^{-6} torr). The capsule was made with a flat, slim base to obtain the best possible contact with the thermocouple/sample-holder of the apparatus and to minimize discontinuity in the thermal exchange between sample and thermocouple. As an inert reference, alumina powder (Al_2O_3), also vacuum-encapsulated, was used.

A Gumbel distribution function was used to analyze the DTA thermograms. This allowed the separation of the DTA peaks corresponding to simultaneous thermal processes and the obtaining of the fusion enthalpies by means of the calculation of the area of the peaks.

3. Results and Discussion

In the ingot photomicrograph (Figure 4) second-phase clusters, characterized by the contrast difference by atomic number, can be observed when the backscattered electron detector is used. These clusters had two morphologies: one spherical and the other in the form of rods. Sizes varied between 1 μm and 50 μm in length, measured using the scale provided by the software of the equipment.

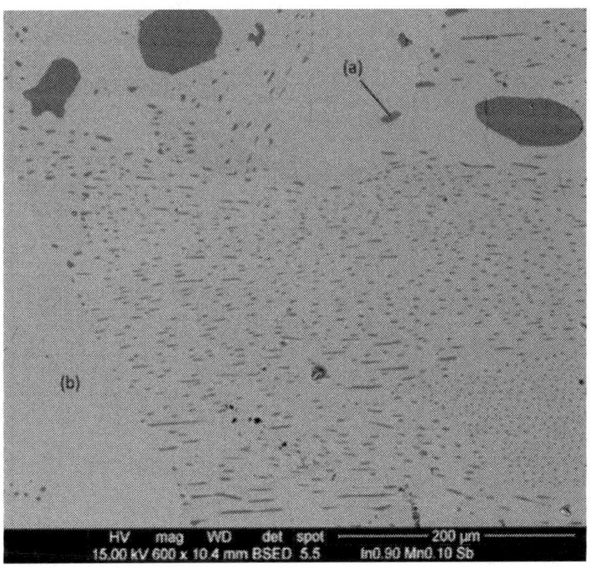

Figure 4. Photomicrograph of a section of the InSb:Mn/MnSb ingot. The manganese antimonide clusters are in the dark areas (a) while the Mn-doped indium antimonide matrix is the light area (b).

Composition of the clusters, as well as that of the surrounding matrix, was determined semi quantitatively using EDS. Figure 5 shows the EDS spectra for the different ingot regions denoted in Figure 4 by (a) and (b). The cluster phase showed a Mn/Sb = 0.896 ± 0.025 atomic ratio, while the indium antimonide matrix showed Mn in low concentration and an In/Sb ratio = 1.013 ± 0.005. The presence of the carbon peak is most likely due to contamination inside the chamber or in the base used for the sample although the possibility that it comes from the graphite of the capsule for synthesis cannot be ruled out, and the oxygen peak indicates a slight oxidation of the sample due to exposure to air.

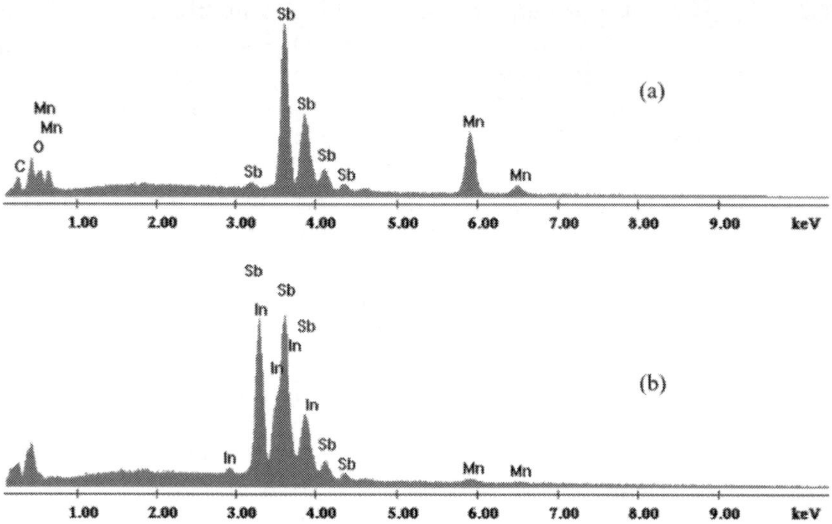

Figure 5. Energy-dispersive X-ray spectroscopy for (a) manganese antimonide clusters and (b) Mn-doped indium antimonide matrix. Intensity is displayed in arbitrary units.

In the X-ray powder diffraction pattern (Figure 6) analysis, the WinPLOTR program [12] was used to read the 2θ positions of the peaks and obtain the Miller indexes, the crystalline structure and the merit figures.

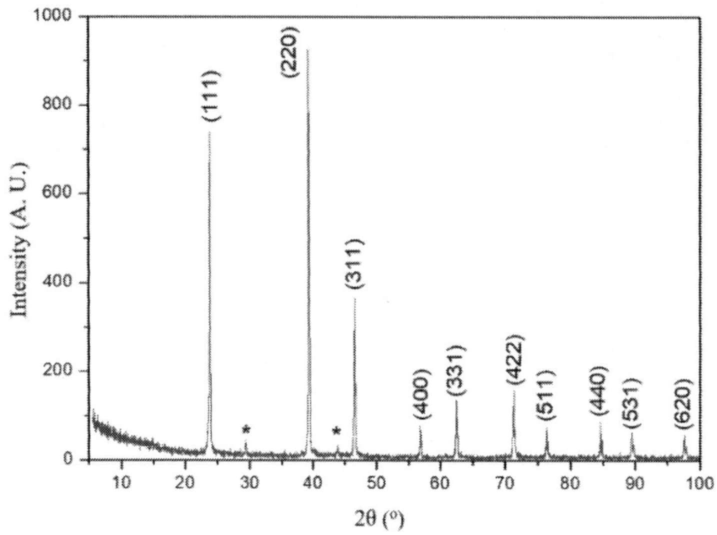

Figure 6. Diffractogram showing the Miller indices (*hkl*) of the reflections of the Mn-doped InSb compound in the InSb:Mn/MnSb ingot. The peaks marked with * correspond to Sb-rich MnSb.

The indexation showed a majority phase with zincblende (ZnS) type cubic structure. Figure 6 shows the Miller indices for the indexed reflections, corresponding to the Mn-doped InSb. The figure of merit $M\ (10) = 680$ was obtained. The lattice parameter $a = 6.474173 \pm 0.000037$ Å and the unit cell volume $V = 271.36$ Å^3 obtained, are slightly lower than the values $a = 6.47937$Å and $V = 272.02$ Å^3 reported for the binary InSb [6], and according to the trend shown by the calculated lattice parameters [7] for the $In_{1-x}Mn_xSb$ system, in which manganese is incorporated into the lattice with an ionic radius lower than that of the indium. Non-indexed reflections are attributed to the antimony-rich MnSb [8]. Table 2 summarizes all obtained values. Lattice parameters and figures of merit as obtained from WinPLOTR are shown in Table 2 footnotes.

Table 2. Indexing of the X-ray powder diffraction pattern for Mn-doped InSb in the InSb:Mn/MnSb ingot.

No.	H K L	SST-OBS	SST-CALC	DELTA	2TH-OBS	2TH-CALC	D-OBS
1	1 1 1	0.042468	0.042469	-0.000001	23.785	23.785	3.7379
2	2 2 0	0.113249	0.113251	-0.000001	39.331	39.331	2.2890
3	3 1 1	0.155733	0.155720	0.000013	46.486	46.484	1.9520
4	4 0 0	0.226523	0.226502	0.000021	56.842	56.839	1.6185
5	3 3 1	0.268993	0.268971	0.000022	62.483	62.480	1.4852
6	4 2 2	0.339758	0.339752	0.000006	71.308	71.307	1.3215
7	5 1 1	0.382209	0.382221	-0.000013	76.374	76.376	1.2460
8	4 4 0	0.452997	0.453003	-0.000006	84.606	84.607	1.1445
9	5 3 1	0.495466	0.495472	-0.000006	89.480	89.481	1.0943
10	6 2 0	0.566246	0.566254	-0.000008	97.614	97.615	1.0237

TOTAL NUMBER OF LINES = 10, NUMBER OF LINES OBSERVED = 10
NUMBER OF LINES CALCULATED = 10, LINES NOT INDEXED = 0
a = 6.474173 ± 0.000037 Å α = 90.000000 ± 0.000000 DEG
b = 6.474173 ± 0.000037 Å β = 90.000000 ± 0.000000 DEG
c = 6.474173 ± 0.000037 Å γ = 90.000000 ± 0.000000 DEG
Unit Cell Volume = 271.36 Å^3
M(10) = 680 AV. EPS. = 0.0000097 F 10 = 181. (0.001288, 43)
M CF. J. APPL. CRYST. 1(1968)108 F CF. J. APPL. CRYST. 12(1979)60

DTA runs were performed at 10 and 20 °C/min. As an example, Figure 7 shows the thermogram obtained at 10 °C/min. Although the DTA technique has been used to measure transitions and magnetic properties [13-15], the thermograms showed no phase transitions (peaks) between room temperature and 400 °C. The reported magnetic transitions below 400 °C in the antimony-rich side of the phase diagram for the binary Mn-Sb system [10] were thus not observed, probably due to the low MnSb concentration of the sample.

On the other hand, in both thermograms two heating transitions were observed, practically simultaneous and identified thanks to the derivative of the DTA signal, and two transitions in cooling.

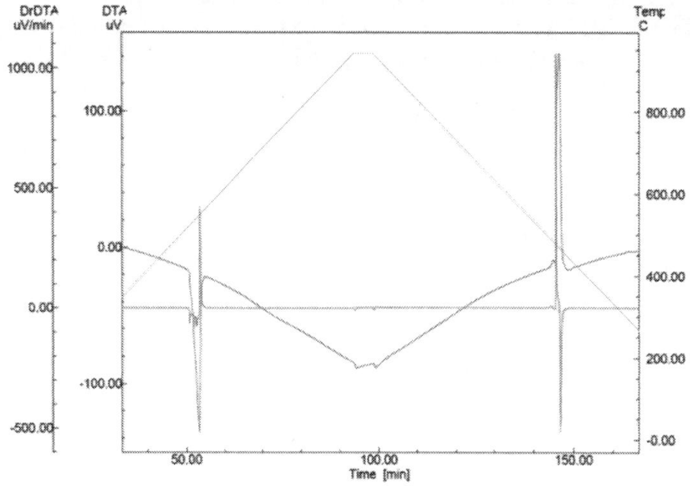

Figure 7. Thermogram obtained by DTA of a sample of the InSb:Mn/MnSb ingot at 10 °C/min. The red line represents the DTA signal; violet, the first derivative of the signal; and green, temperature.

The minimums (heating) and maximums (cooling) of the derivative of the DTA signal are associated with the temperatures at which the velocity of change of the signal is greater (onset); that is, with the phase transition temperatures.

Temperatures associated with the phase transitions at each heating/cooling rate, as well as the corresponding extrapolated temperatures at 0 °C/min [16] are shown in Table 3 and in Figure 8.

The transition with highest signal intensity is associated with the fusion of the InSb:Mn matrix between 485 °C and 528 °C (black in Figure 8), in contrast to the congruent fusion of the binary InSb at 527.7 °C [6]. The presence of Mn and of the MnSb clusters thus lowers the thermal stability of the matrix compound.

Table 3. Phase transition temperatures and extrapolation at 0 °C/min for a sample of the InSb:Mn/MnSb ingot.

	Temperatures in Heating (°C)		Temperatures in Cooling (°C)	
10 °C/min	520	538	490	478
20 °C/min	531	548	486	471
Extrapolation at 0 °C/min	509	528	494	485

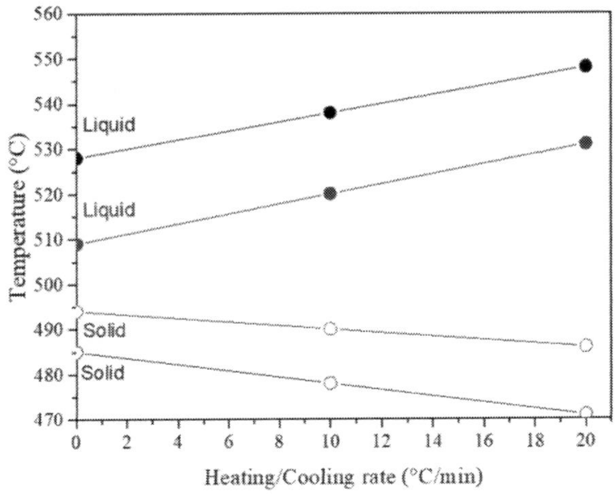

Figure 8. Extrapolation at 0 °C/min of the phase transition temperatures obtained by DTA for Mn-doped InSb (black) and antimony-rich MnSb (blue). Filled markers represent heating temperatures and open markers represent cooling temperatures. Solid lines show the linear adjustment in each case.

Between 494 °C and 509 °C (blue in Figure 8) the transition with lower signal intensity takes place. This is associated with the fusion of antimony-rich manganese antimonide clusters (Mn/Sb = 0,896). These temperatures are lower than those reported starting at 570 °C for the antimony-rich compositions of the Mn-Sb binary system phase diagram [10].

Enthalpy is a thermodynamic property that allows to estimate heat transferred during a constant pressure process. Semiconductor manufacturing includes constant pressure growth techniques in the absence of reagents and gaseous products. This means reaction-associated work is minimal and the change in enthalpy can approximate the energy involved in the process. Knowledge of the thermodynamic properties of a compound, including the enthalpy of phase transformation, is thus important in making good quality crystals and thin films on a large scale.

The enthalpies of phase transformation, ΔH, can be obtained using the Borchardt-Daniels model [17], according to which in the course of a reaction or transformation:

$$\Delta H = -K \int_{t=0}^{t=\infty} \Delta T \mathrm{d}t = -KA \tag{1}$$

as far as there is no change in the heat capacity C of the sample during the transformation and the DTA responses ΔT are equal before and after the transformation. Here, K is the value of the DTA device calibration curve at the temperature associated with a peak, and A, the area contained in that DTA signal peak. The calibration procedure of the DTA apparatus, using pure metals (under the same run conditions as the samples) with known fusion enthalpies, was described by Matsushita et al [16]. For the device used in this work, the best fit (R^2 = 0.995) for the calibration points resulted of the form:

$$K(T) = -0.00157 + 0.00175 e^{\left(\frac{T}{2953.27}\right)} \text{ (cal/}\mu\text{V}\cdot\text{s)} \tag{2}$$

However, the determination of the area A of the peak, and therefore the evaluation of ΔH, can be affected by a series of experimental conditions that can be sources for error. A fairly frequent situation is a signal baseline displacement due to a change in the heat capacity of the sample during the transition. To correct this, the baseline must be subtracted prior to carrying out integration. Given that this baseline is not straight, a method must be developed that allows the ΔH value to be obtained leaving out the errors resulting from the fact that $\Delta C \neq 0$.

Clavaguera et al [18] developed an accurate method for DSC (Differential Scanning Calorimetry) signal processing using an iterative program to calculate ΔH from the decomposition of the signal in

Journal of Physics: Conference Series **2238** (2022) 012004 doi:10.1088/1742-6596/2238/1/012004

the input of the transition itself and in the input due to ΔC. With the adaptation of this method to DTA signals [19] the value of ΔH is given directly by the equation:

$$\Delta H = \int_{t=0}^{t=\infty} [K(|\Delta T_t| - |\Delta T_{t=0}|) - \beta f \Delta C] \mathrm{d}t \tag{3}$$

where β is the heating rate and f is the sample transformed fraction. The area between two curves must, therefore, be obtained with an iterative program. The iteration process (see Figure 9) to obtain the area between the peak and the corrected baseline begins with indicating the start point (y_0, t_0) and the end point (y_f, t_f) to the program, which then calculates h_∞. A straight line is defined between both points, on the basis of which total area and partial areas are calculated, and with these the transformed fraction f at each instant t_i (see Figure 9). The set of points ($f_i h_\infty$, t_i) thus generated constitutes a new baseline. The calculation is repeated until variation in the total area is less than 0.01%.

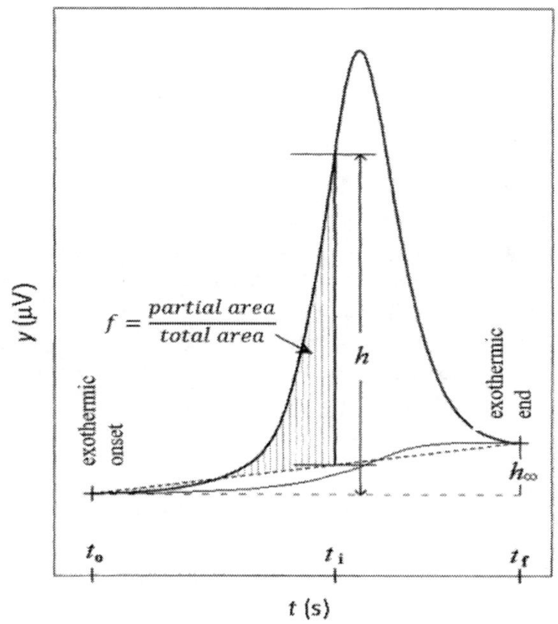

Figure 9. Diagram of a DTA curve, including a solidification peak (exothermic) with change in the heat capacity, indicating the subtended area and the transformed fraction f for a particular time t_i [19].

For the thermograms obtained in this work, the peaks must first be separated before using the iteration process. Peak adjustment was carried out empirically by means of a Gumbel function [20]. This family of distributions, which is normally used to model extreme values, is characterized by its asymmetry: it is a distribution with an elongated rightward (or leftward) tail used to describe maximum or minimum values. The probability density function is given by:

$$y(x) = y_0 + A\,exp\left\{-\,exp\left[-\left(\frac{x-x_c}{\omega}\right)\right] - \left(\frac{x-x_c}{\omega}\right) + 1\right\} \tag{4}$$

where the parameter x_c defines the center of distribution, and therefore its asymmetry. The aforementioned asymmetry is, precisely, the characteristic that combines with the DTA peaks. These are poorly adjusted by traditional distributions (such as Gaussian, for example) that are normally symmetric in relation to the mean value.

To verify its viability, the Gumbel function was initially applied to individual DTA heating peaks, and satisfactory adjustments ($R^2 > 0.98$) were obtained. In its application to the deconvolution of DTA

 2238 (2022) 012004 doi:10.1088/1742-6596/2238/1/012004

peaks superimposed due to simultaneous transitions, such as those presented in Figure 7, correlation coefficients for adjustments were greater than 0.99.

Figures 10 and 11 summarize the peak separation process and the area calculation between the DTA signals and the corrected baselines.

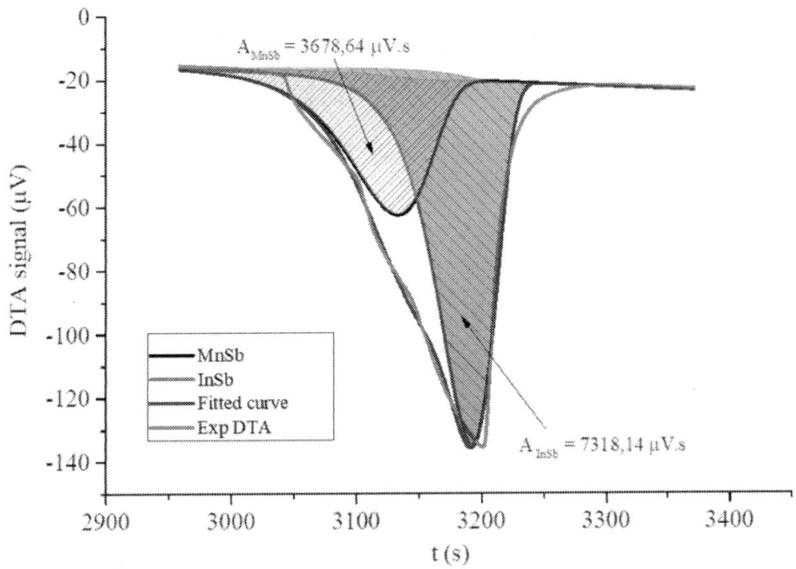

Figure 10. Peak separation and calculation of areas in the heating of InSb:Mn/MnSb at 10 °C/min.

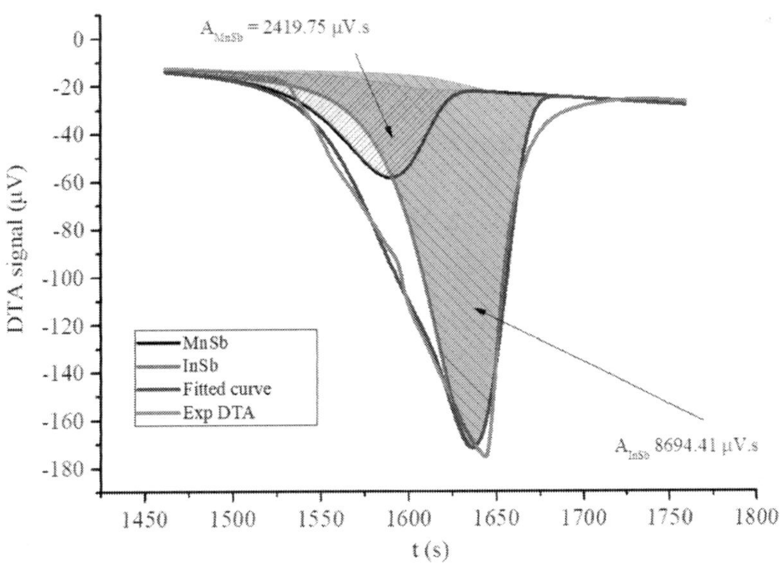

Figure 11. Peak separation and calculation of areas in the heating of InSb:Mn/MnSb at 20 °C/min.

The enthalpies of fusion can be calculated using the results shown in Figures 10 and 11 and Equation 2 to find the *K* associated with the temperature of each peak. The number of moles of each compound can be obtained through stoichiometric calculations using the weight of each element used for the synthesis, their atomic weights and the amount of sample used for the DTA. All this is shown in Table 4.

Table 4. Fusion enthalpies ΔH_f for InSb:Mn and MnSb.

	MnSb	InSb:Mn
$A_{Average}$ (μV·s)	3049.20	8006.28
K (cal/μV·s)	K(509 °C) = $5.09 \cdot 10^{-4}$	K(528 °C) = $5.23 \cdot 10^{-4}$
$KA_{Average}$ (cal)	1.55	4.19
Moles	$3.57 \cdot 10^{-5}$	$3.21 \cdot 10^{-4}$
ΔH_f (Kcal/mol)	117.4	4.8

The enthalpy of fusion found for InSb:Mn is lower than the 11.2 Kcal/mol reported for the binary InSb [21], which confirms once again that the incorporation of Mn subtracts thermal stability from indium antimonide. On the other hand, the enthalpy of fusion for antimony-rich MnSb is much greater than that of InSb:Mn. The enthalpy of fusion for binary MnSb was not found in the literature.

4. Conclusion

Synthesis of $In_{0.9}Mn_{0.1}Sb$ by direct fusion of the constituent elements, as well as by other techniques reported in the literature, resulted in a polycrystalline ingot with *p*-type conductivity, made up of antimony-rich manganese antimonide clusters in a manganese-doped indium antimonide matrix. The Mn-doped InSb matrix showed a cubic structure with lattice parameter $a = 6.474173$ Å and unit cell volume $V = 271.36$ Å^3; lower than the values reported for the InSb binary. It was found that the fusion of Mn-doped InSb matrix occurs between 485 °C and 528 °C, while the fusion of the Sb-rich MnSb clusters occurs between 494 °C and 509 °C, showing lower thermal stability than the InSb and MnSb binaries, respectively. The enthalpies of fusion were estimated at 4.8 Kcal/mol for InSb:Mn and 117.4 Kcal/mol for antimony-rich MnSb.

Acknowledgments
This work was funded by the Research Direction of the Pontificia Universidad Católica del Ecuador project O13121 and by CONDES-LUZ project No. CC-0146-14.

References

[1] Gupta A 2004 *Novel room temperature ferromagnetic semiconductors* (Doctoral Dissertation, Royal Institute of Technology, Sweden) pp 1-7 Available: https://cloudfront.escholarship.org/dist/prd/content/qt352984v3/qt352984v3.pdf
[2] Kumar S, Koo BH, Sharma SK, Knobel M and Lee CG 2010 *Nano* **5** 349-55
[3] Ohno H 1998 *Science* **281** 951-6
[4] Jungwirth T, Sinova J and MacDonald AH 2003 *A. Phys. Pol. A* **104** 103-12
[5] Liu C, Li Y and Zeng Y 2010 *Engineering* **2** 617-24
[6] Zhang W and Razeghi M 2004 *Antimony-based materials for electro-optics* (In: T. Steiner (Ed), *Semiconductor nanostructures for optoelectronic applications*. Artech House Inc., Norwood, MA USA) pp 242-50
[7] Verma UP, Devi N, Sharma S and Jensen P 2011 *Eur. Phys. J. B* **81** 381-6
[8] Novotortsev VM, Kochura AV, Marenkin SF, Fedorchenko IV, Drogunov SV, Lashkul A and Lähderanta E 2011 *Russ. J. Inorg. Chem.* **56** 1951-6
[9] Pashkova ON, Sanygin VP, Ivanov VA, Padalko AG and Novotortsev VM 2006 *Inorg. Mater.* **42** 459-62

[10] Sanygin VP, Pashkova ON, Filatov AV and Novotortsev VM 2010 *Inorg. Mater.* **46** 807-11
[11] Rednic L, Deac IG, Dorolti E, Coldea M, Rednic V and Neumann M 2010 *Cent. Eur. J. Phys.* **8** 620-7
[12] Roisnel T and Rodriguez-Carvajal J 2001 *Mater. Sci. Forum* **378-381** 118-23
[13] Vollstädt H 1968 *Geophysical Journal International* **16** 71-7
[14] Sidorov VA 1998 *Appl. Phys. Lett.* **72** 2174-5
[15] Rotstein Habarnau Y, Bergamasco P, Sacanell J, Leyva G, Albornoz C and Quintero M 2012 *Physica B: Condensed Matter* **407** 3305-7
[16] Matsushita H, Endo S and Irie T 1991 *Jpn. J. Appl. Phys.* **30** 1181-5
[17] Borchardt HJ and Daniels F 1957 *J. Am. Chem. Soc.* **79** 41-6
[18] Clavaguera N, Clavaguera-Mora MT and Fontana M 1998 *J. Mater. Res.* **13** 744-53
[19] Durante Rincón CA 1998 *Crecimiento, caracterización y propiedades ópticas de los sistemas* $CuIn_xGa_{1-x}Se_2$ *y* $CuIn_xAl_{1-x}Se_2$ (Doctoral Dissertation. Universidad Autónoma de Barcelona. Bellaterra, España) pp 209-214
[20] Gumbel EJ 1960a *Bull. Inst. Internat. de Statistique* **37** 471-5
[21] Nachtrieb NH and Clement N 1958 *J. Phys. Chem.* **62** 876-7

Journal of Physics: Conference Series **2238** (2022) 012005 doi:10.1088/1742-6596/2238/1/012005

Environmentally friendly synthesis of silicon dioxide nanoparticles and their application for the removal of emerging contaminants in aqueous media

Jennifer Tejedor[1], Victor H. Guerrero[1,*], Karla Vizuete[2] and Alexis Debut[2]

[1]Department of Materials, Escuela Politécnica Nacional, Quito 170525, Ecuador
[2]Centro de Nanociencia y Nanotecnología, Universidad de las Fuerzas Armadas (ESPE), Sangolquí 171103, Ecuador

[*]Corresponding author: victor.guerrero@epn.edu.ec

Abstract. The highly detrimental potential effects that emerging contaminants have for the environment and human health demand the development of sustainable technologies to eliminate these pollutants. In this work, an eco-friendly approach was followed to synthesize SiO_2 nanoparticles to be used to remove caffeine and triclosan from synthetic waters. Rice husk was the precursor used, which was subjected to acid leaching treatments using acetic and oxalic acids, and pyrolysis for 2 h at 700 °C. The resulting powders were characterized by X-ray diffraction, Raman and infrared spectroscopies, the Brunauer-Emmett-Teller method, and scanning and transmission electron microscopy. The particles synthesized were amorphous, had an average size between 4.6 and 9.6 nm, and specific surface areas between 208 and 223 m^2/g. The adsorptive performance of the nanosilica obtained using acetic acid was studied by batch tests. The optimum dosage, maximum removal efficiencies and maximum adsorption capacities were 8 g/L, 76%, and 2.74 mg/g, and 24 g/L, 48%, and 0.75 mg/g for triclosan and caffeine, respectively. The data collected fitted the Sips isotherm model, and the pseudo-second and -first order kinetics models, for triclosan and caffeine, respectively. The promising results obtained open the possibility of using the synthesized nanomaterials for removing a variety of toxic and recalcitrant pollutants.

1. Introduction

Emerging contaminants (ECs) are chemical compounds recently detected in aquatic ecosystems and wastewater. Due to the lack of monitoring and regulations, these pollutants are continuously introduced in environmental matrices and their persistence represents a threat to ecological and human welfare [1,2] Pharmaceuticals, personal care, agrochemical products, metabolites, food additives, and hormones are among these contaminants [3]. Particularly, triclosan (TCS) is a bactericidal widely used in personal care products, cosmetics, paints and plastic additives [4]. It is an antimicrobial compound with high hydrophobicity ($\log_{kow} = 4.8$), low water solubility (2000 - 4600 µg/L at 25 °C) [5] and it is highly toxic for the environment. TCS is considered an endocrine disruptor (i.e. affects fecundity and fertility) and produces teratogenic effects (i.e. embryo mortality and hatchability) [6]. Caffeine (CAF) is ubiquitously found in water bodies, and it is a stimulant of the nervous system present in drinks, food and even medicines [7]. CAF is an anthropic indicator of contamination due to its regular consumption (70 - 400 mg/person/day) and easy mobility in the environment (solubility of 20 g/L) [1]. Several studies have shown that CAF is highly toxic for aquatic organisms, including for instance cladocerans (*Ceriodaphnia dubia*: 48 h-LC_{50}= 60 mg/L) and fish (*Pimephales promelas*: 48 h-LC_{50}= 100 mg/L) [8].

Given their potentially adverse effects, it is critical to develop sustainable alternatives and efficient methods for the removal of ECs, including eco-friendly processes, using low or no cost materials (recovery of waste), and avoiding secondary pollution [9]. Faced with this perspective, one of the main challenges for wastewater treatment is the design of materials capable of retaining contaminants. In this regard, nanotechnology has made it possible to obtain emerging materials whose characteristics have proven to be efficient in immobilizing water pollutants [10,11]. An interesting alternative is silicon dioxide nanoparticles (NPs) since they exhibit high specific surface area, adjustable pore size and ability to enclose hydrophilic and hydrophobic pollutants [12].

Silica NPs have been synthesized through different routes. The most common techniques include Stöber method [13], micro emulsion [14], chemical vapor condensation [15], mechanical attrition [16] and sol-gel method [17]. Nevertheless, these methods have many limitations such as high energy consumption associated with high temperature, high pressure and strong acidity, or the use of expensive chemical precursors. Therefore, these approaches to synthesize NPs may not be sustainable [18]. Accordingly, a green synthesis approach has been promoted in the last decades, focused on the use of biomass as a natural source for the assembly of nanoparticles, whose application represents an attractive alternative, especially when considering taking it to a commercial scale [19,20]. Previous studies highlight the feasibility of production of silica NPs from rice husk (RH) and their potential as adsorbent of ECs such as dyes, pharmaceutical products, heavy metals, etc. For instance, silica NPs have been used for fluoride removal, eliminating 72% in 60 min, with 4 g/L of the adsorbent [21]. They have also been used for removing methylene blue, where 95% of the removal was achieved in 30 min [22]. On the other hand, silica loaded with iron oxide has been used as an arsenic adsorbent, obtaining an efficiency of 97.5% using 0.1 g/L of adsorbent with a contact time of 90 min [23].

In this work, silica NPs were synthesized through an innovative and eco-friendly approach using RH as a natural precursor. The NPs obtained were characterized and used as adsorbent for removing two common ECs, caffeine and triclosan. The optimization of different factors, including the adsorbent dose and contact time was carried out. Additionally, kinetic and isotherm studies were performed.

2. Methods

2.1. Materials and reagents

The RH was obtained from a rice grinding company situated in Babahoyo city (location S 2°11′23.1", W 79° 31′ 60") in Los Ríos province, in Ecuador. The residues were washed with deionized water, and dried for 24 h in an oven at 60°C. A Thomas Wiley, model 3379-K05, blade mill was used to grind the RH, which was then sieved to obtain particles with sizes between 250 and 355 μm [24].

Chemicals were reagent grade and not subject to any purification process. Acetic acid (glacial) (100% pure) was purchased from Merk KGaA, hydrochloric acid (HCl, 37%) from Pharmco-Aaper, oxalic acid (> 99.5%) from Sigma Aldrich, and caffeine (99%) and triclosan (99%) from Thermo Fisher Scientific.

2.2. Synthesis of SiO_2 nanoparticles

The nanosilica was synthesized through a thermal method that followed an acid leaching pretreatment for impurity removal [25]. Three different approaches were adopted for the chemical pretreatment of RH, which consisted on using the conventional inorganic acid HCl, and two organic acids: acetic acid (AA) and oxalic acid (OA). For the first and second pretreatments, RH was added to beakers containing 10 w/w % HCl and 5 w/w % acetic acid solutions (1:10 g/mL ratio of rice husk to acid), respectively. The solutions were heated up to 95 ± 5 °C and stirred continuously for 2 h. The resulting samples will be referred to as RH-HCl and RH-AA. For the third approach, the acid leaching of RH was performed using a 0.5 M oxalic acid solution (RH-OA) under subcritical water conditions (50 bar and 155° C) for 2 h in a high-pressure reactor (Berghof, BR-500).

The RH-HCl, RH-AA, and RH-OA samples were filtered, washed to remove the acid residues, dried at 100 °C for 24 h, and calcined at 700 °C for 2 h in a furnace (Carbolite, CWF 1300) operating with a 5 °C/min heating rate. The pyrolyzed samples were finally ground until white silica powders were obtained. These nanopowders will be referred to as SiO_2-HCl, SiO_2-AA, and SiO_2-OA.

2.3. Characterization

RH was characterized according to ASTM standards, to determine their lignin, hemicellulose and cellulose (D1106-21 and D1109-21), extractives (D1107-21 and D1110-21), ash (E1755-01), and moisture content (D4442-20). X-ray fluorescence (XRF) (Bruker AXS, S8 Tiger) was used to determine the chemical composition of RH ash. X-ray diffraction (XRD) analyses of the nanopowders were carried out on a Bruker D2 phaser instrument with Cu Kα (1.54184 Å) radiation at 30 kV and 10 mA, over a 2θ range of 10 – 70°. The specific surface area, the pore size distribution, and the adsorption and desorption isotherms of the powders were determined by the Brunauer-Emmett-Teller method (BET), employing a NOVA touch 1LX instrument. Samples were degassed before measurements at 200 °C for 720 min. The functional surface groups were identified by Fourier-transform infrared (FT-IR) and Raman spectroscopies. A Perkin Elmer Spectrum 100 instrument was used to record infrared spectra between 500 and 4000 cm^{-1}. Raman spectra were registered between 100 to 2000 cm^{-1} using a Horiba Scientific LabRam Evolution instrument, at a wavelength λ= 532 nm. The scanning electron microscopy (SEM) images were obtained using a Tescan instrument with a Shottky field emitter. For these analyses, an aliquot of the material was fixed to an aluminum sample holder using a double layer of carbon tape and then samples were analyzed. The transmission electron microscopy (TEM) analyses were performed using a FEI Tecnai G2 Spirit Twin microscope. The samples were prepared as follows; first, materials were homogenized in an ultrasonic bath for 10 min. After that, a drop of solution was placed on the formvar/carbon mesh, aqueous media was absorbed using filter paper and the samples were measured.

2.4. Adsorption experiments

Adsorption batch tests were carried out and the optimal adsorbent dosage and contact time were determined. The SiO_2-AA NPs were considered a good alternative for the adsorption test because of their high surface area. The adsorbent dosages investigated were over a range of 0.25 to 36 g/L. The SiO_2-AA NPs were added to 10 mL of aqueous solutions of the contaminants. The solutions had concentrations of 10 mg/L and were sonicated for 1 min and then stirred at 150 rpm (Orbital Shaker Tecnal, TE-1400) for different periods of time (0 – 150 min) at room temperature (20 ± 3 °C). The CAF and TCS solutions were filtered and analyzed using a UV-VIS spectrophotometer (Horiba Scientific, Duetta model) at 272 and 292 nm, respectively. The adsorptive capacities (q_e, mg/g) were determined by equation (1):

$$q_e = \frac{(C_o - C_e)\ v}{m} \tag{1}$$

where C_o and C_e (mg/L) are the initial and equilibrium concentrations. The volume of the aqueous solutions is v (L), and the mass of SiO_2 NPs is m (g).

2.5. Kinetics studies and adsorption isotherms

The optimal adsorbent dosage and contact time were used for the kinetics and equilibrium studies, as well as the same conditions for the equilibrium studies (20 ± 3 °C, stirring at 150 rpm). The initial concentration for adsorption kinetics was fixed to 10 mg/L, and times between 0 and 150 min were considered. For the isotherm studies, the concentration of the contaminant was between 1 and 50 mg/L. The data collected were analyzed through the pseudo-first and second order non-linear models (Eq. 2 and Eq. 3, respectively), and the intra-particle diffusion (IPD) kinetics model (Eq. 4).

$$q_t = q_e(1 - e^{-k_1 t}) \tag{2}$$

$$q_t = q_e \frac{q_e k_2 t}{1 + q_e k_2 t} \tag{3}$$

$$q_t = k_p\sqrt{t} + C \tag{4}$$

where q_t and q_e (mg/g) are the masses of contaminant adsorbed per gram of SiO_2 NPs after time t and at equilibrium, respectively. The rate constants for the pseudo-first order, pseudo-second order, and the IPD kinetics models are k_1 (min^{-1}), k_2 (g/mg min) and k_p (mg/g min$^{1/2}$). The constant C (mg/g) reflects the boundary layer effect [26,27].

Furthermore, the data collected were fitted to the Langmuir, Freundlich, Sips, and Temkin non-linear models (equations 5, 6, 7 and 8, respectively). In these models, the equilibrium adsorptive capacity q_e (mg/g) is:

$$q_e = \frac{q_{\max} K_L C_e}{1+ K_L C_e} \tag{5}$$

$$q_e = K_f C_e^{\frac{1}{n}} \tag{6}$$

$$q_e = \frac{q_m (K_s C_e)^{1/m}}{1+(K_s C_e)^{1/m}} \tag{7}$$

$$q_e = \frac{RT}{b}\ln(A_T C_e) \tag{8}$$

where q_{max} (mg/g) is the maximum adsorptive capacity. K_L (L/mg) is the constant of the Langmuir equation, C_e (mg/L) is the contaminant concentration at equilibrium, K_f [(mg/g) (L/mg) $^{1/n}$] is the constant of Freundlich, n is a constant associated with adsorption intensity. q_m (mg/g) is the adsorption capacity, K_S (L/mg) is the energy adsorption constant, and m is the exponent (system heterogeneity) for the Sips model. A_T (L/mg) is the maximum energy adsorption constant for the Temkin model, b (J/mol) is the heat of sorption. R (J/mol K) and T (K) are the gas constant and the absolute temperature, respectively [1,27].

3. Results and discussion

3.1. Characterization

RH is considered a lignocellulosic residue because more than 75% of the dry weight corresponds to organic compounds (lignin, hemicellulose and cellulose), followed by a remaining 20% of inorganic component (ash) [28]. The composition of the rice husk used in this work is summarized in Table 1.

Table 1. Physico-chemical composition of RH and RH ash, in wt.%.

Parameter (wt.%)	Rice husk	Parameter (wt.%)*	Rice husk ash
Lignin	17.44 ± 0.46	SiO_2	93.51
Hemicellulose	20.03 ± 0.85	CaO	1.05
Cellulose	31.15 ± 1.18	Al_2O_3	0.94
Extractives (ethanol-toluene)	1.41 ± 0.09	MgO	0.56
Extractives (water)	7.22 ± 0.47	K_2O	0.45
Total extractives	8.62 ± 0.39	Fe_2O_3	0.41
Moisture	4.71 ± 0.10	P_2O_5	0.34
Ash*	**22.76 ± 0.25**	Na_2O	0.24

The percentages of lignin, cellulose, hemicellulose and ash were 17.44%, 31.15%, 20.03% and 22.76%, respectively. Furthermore, the XRF results show that SiO_2 is a rice husk ash principal component. Silica represents 93.51% of the total while the residual 4% includes CaO, Al_2O_3, K_2O, MgO, Fe_2O_3, Na_2O and P_2O_5 in low concentrations. After calcination, the yield of nanoparticles was around 0.216 g per gram of rice husk. Thus, rice husk has been used as a source of silica.

The XRD patterns for the SiO_2 NPs synthesized using the different acid pre-treatments examined, SiO_2-HCl, SiO_2-AA, and SiO_2-OA, are shown in Figure 1. The diffractograms exhibited a broad intense peak around of 2θ ≈ 21-23°, which confirmed that the synthesized NPs were amorphous [19]. This indicates that, regardless of the pre-treatment, the calcination process transformed the crystalline cellulose structure in amorphous particles [29]. This is consistent with the fact that crystalline structures are produced at temperatures above 700 °C [30].

Figure 2 shows the Raman spectra for the silica NPs. The peaks between 485 and 610 cm^{-1} are attributed to the Raman spectra of SiO_2 [31]. The stretch band around 980 cm^{-1} belongs to the silanol (Si-OH) band [32]. Finally, hydroxyl (OH) bending bands were detected at 1623, 1626 and 1631 cm^{-1} [33].

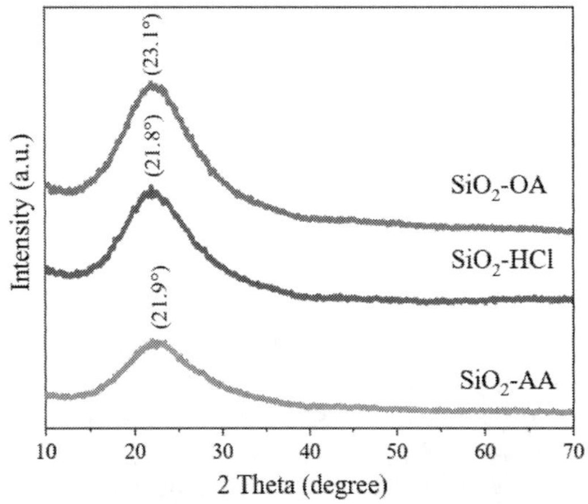

Figure 1. XRD patterns of the SiO_2 NPs.

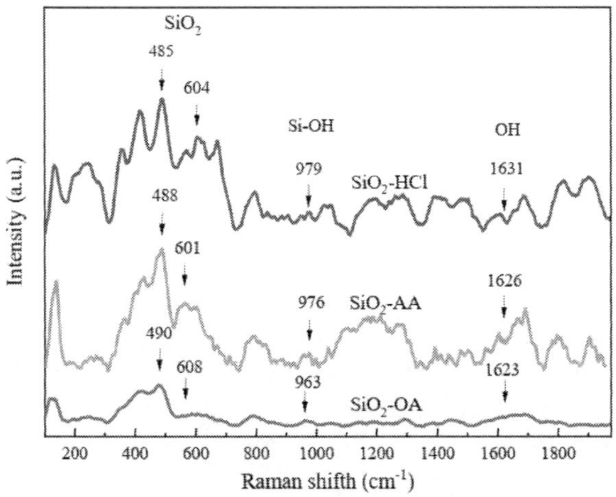

Figure 2. Raman spectra of the SiO_2 NPs.

Figure 3 exhibits the FT-IR spectra obtained for the SiO2 NPs. The SiO2-HCl, SiO2-AA and SiO2-OA nanopowders show characteristic peaks around 568, 546 and 568 cm^{-1}, respectively, which can be related to Si-O-Si bending vibrations [34]. The bands at 756, 766 and 774 cm^{-1} are assigned to Si-O-Si symmetric stretching bond (silanol) vibrations [35]. Moreover, the silicon dioxide structure SiO_2 is shown in tetrahedron asymmetric vibrational stretching of Si-O-Si, which manifested in peaks at 1172, 1116 and 1086 cm^{-1} [19]. The bands at 1686, 1668 and 1662 cm^{-1} correspond to O-H bending vibrations of adsorbed molecular water [35]. The interaction between the hydroxyl groups of silanol (Si-OH) and water adsorbed on the surface of the NPs is manifested in the stretching vibrations at 3770, 3744, and 3810 cm^{-1} [34]. According to Raman and FTIR data, silica nanoparticles have a large range of functional groups namely hydroxyl, siloxane and silanol groups, which could increase adsorption affinity towards contaminants [36].

The high specific surface areas of the synthesized silica NPs (208-223m^2/g) are shown in Table 2, along with the corresponding pore volume and pore radius. The specific surface areas (S_{BET}) obtained in this work are greater than those reported in other studies. For instance, surface areas of 164 m^2/g [18], 208-218 [37], and 208-212 m^2/g [24] were reported when using inorganic acids such as hydrochloric and sulfuric acids. Furthermore, the use of organic acids in the synthesis of NPs has been reported. Leaching with citric acid resulted in an S_{BET} of 198 m^2/g [37] and 112-625 m^2/g [30].

These results are comparable with the ones obtained for the silica NPs treated with acetic and oxalic acid, which showed a large specific surface area. Added to this, the use of organic acids for acid leaching of rice husk represents an ecofriendly method, considering that reduces the environmental impact of the production of acidic wastes using the conventional inorganic acid treatment [38]. Therefore, SiO_2 NPs synthesized with organic acids and, in particular, SiO_2-AA nanoparticles obtained through a relatively simple preparation method make them a good alternative as adsorbents for pollutant removal. On the other hand, the results of the characterization of the NPs in Table 2 show average pore sizes between 0.977 and 2.772 nm for all treatments. According to the IUPAC classification, these materials are considered microporous (< 2 nm) and mesoporous (2 - 50 nm) [39].

Figure 4a shows the adsorption-desorption isotherm for the SiO_2-AA NPs. This isotherm belongs to the IV category, and the hysteresis loops are H3 type, according to IUPAC. It is typical of mesoporous materials and slit-like porous structures [40]. In this type of system, the saturation of the micropores of the material occurs first ($P/P_o \approx 0.35$), followed by the filling of the mesopores with the formation of multilayers of N_2 [41]. Furthermore, as shown in Figure 4b, the pore size distribution exhibited that most of the pore radius values of SiO_2-AA NPs are distributed between 1.59 to 3.63 nm, verifying the mesoporosity of the sample.

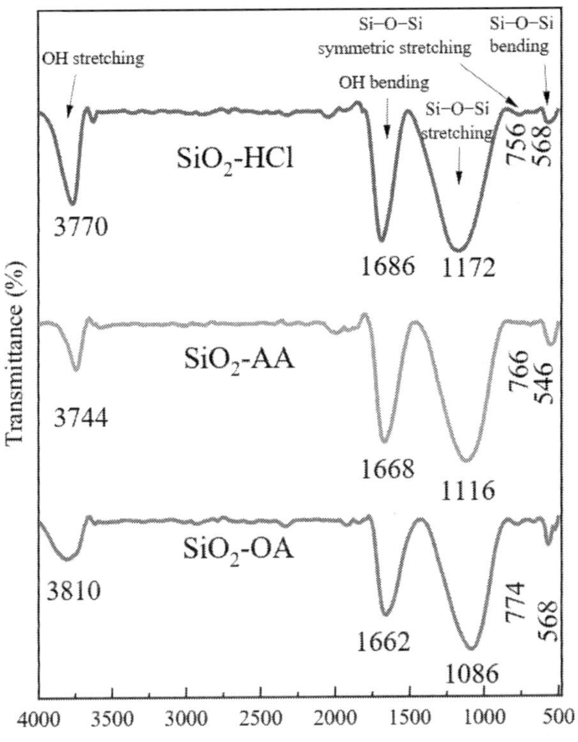

Figure 3. FT-IR spectra for the SiO_2 NPs.

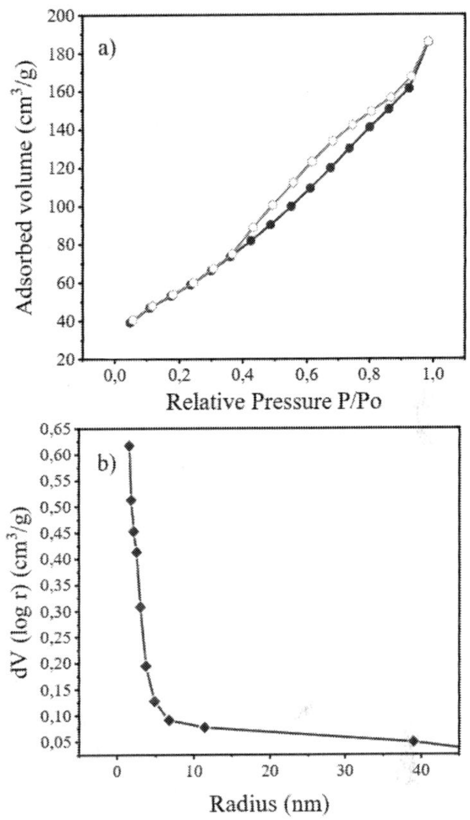

Figure 4. Adsorption-desorption isotherm and pore size distribution for the SiO_2-AA.

Table 2. BET surface area, total pore volume and average pore radius of the silica nanoparticles synthesized from different acid pretreatments.

Sample	S_{BET} (m^2/g)	Total pore volume (cm^3/g)	Average pore radius (nm)
SiO_2-AA	207.73	0.288	2.772
SiO_2-OA	222.97	0.109	0.977
SiO_2-HCl	208.05	0.101	2.641

SEM and TEM images of the silica nanopowders obtained through different acid pretreatments are shown in Figure 5. The SEM micrographs show that the three treatments used produced powders with similar characteristics, composed of nanoparticle agglomerates and a few relatively big porous particles which are calcined residues of the rice husk used as precursor. During the pyrolysis process, the thermal decomposition of the organic matter of RH (lignin, cellulose and hemicellulose) occurs and, as a result, the particles with porous structure are obtained [42]. Nevertheless, most of the sample is composed of agglomerates of nanoparticles. Furthermore, it is common for small-sized particles to aggregate and form larger particles after the calcination at 700 °C [43]. This is also evidenced by the TEM images, which exhibit a strong agglomeration of nanometric-sized particles whose size ranges from approximately 0.95 to 18.75 nm. However, these agglomerates do not seem to substantially affect the specific surface area, since the BET analyses showed values greater than 200 m^2/g. According to the size distribution histograms, the synthesized SiO_2-AA, SiO_2-OA and SiO_2-HCl nanoparticles showed an average particle size of 5.4, 9.6 and 4.6 nm, respectively. Consistently with the XRD patterns previously presented, the SAED patterns included next to the TEM images (Figure 5) showed diffraction rings that suggest an amorphous structure of the material.

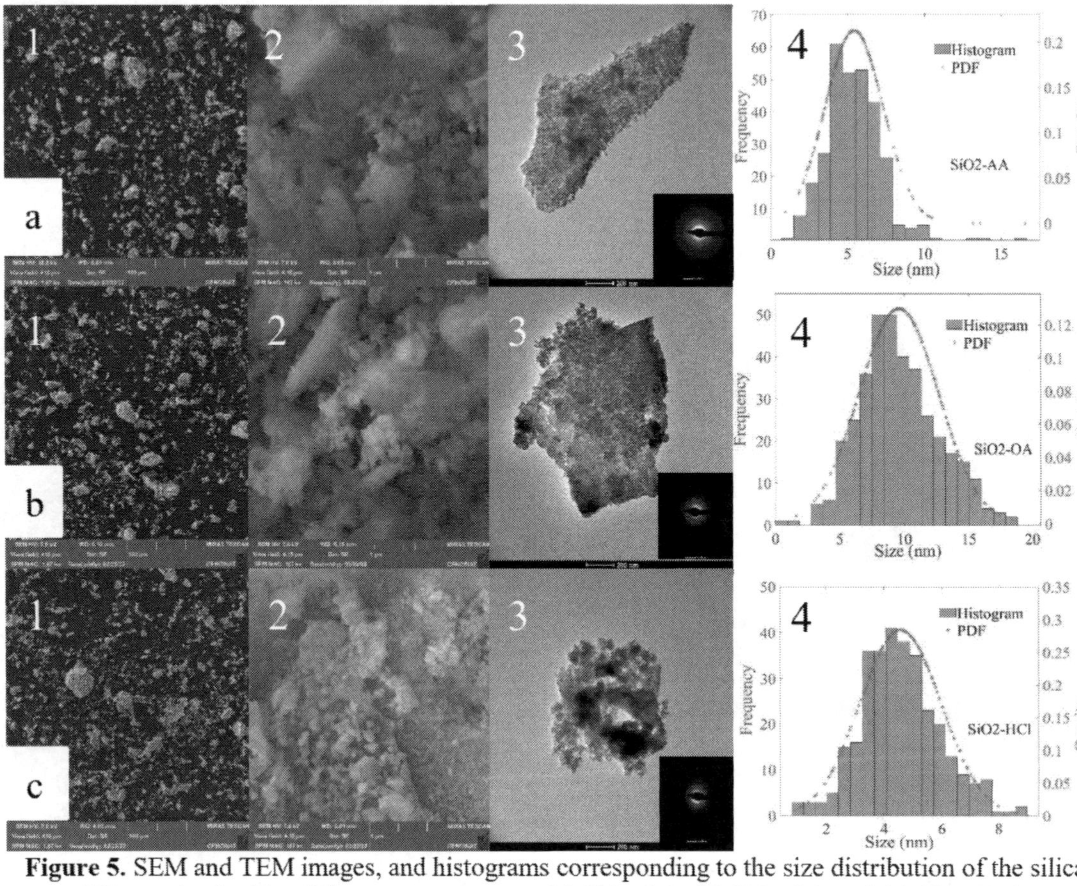

Figure 5. SEM and TEM images, and histograms corresponding to the size distribution of the silica NPs synthesized by different pretreatments (a) SiO_2-AA, (b) SiO_2-OA and (c) SiO_2-HCl.

3.2. Adsorption process

3.2.1. Effect of the dose. The surface area available for adsorption depends on the adsorbent dosage. The NP doses evaluated are in the range of 0.25 to 12 g/L for TCS, and 1 to 36 g/L for CAF. The contact time and initial concentration of the contaminant were fixed at 60 min and 10 mg/L, respectively. According to Figure 6, increasing the dose of NPs from 0.25 to 8 g/L and from 1 to 24 g/L, the TCS and CAF removal efficiencies were increased from 47.86 to 76.18% and from 5.38 to 47.72%, respectively. Therefore, the higher availability of active adsorption sites improves the contaminant adsorption performance [1,44]. Nevertheless, at dosages higher than 8 and 24 g/L, only slight increases in the adsorption efficiency were observed of 78.13% and 50.43% for TCS and CAF, respectively. This could be a result of particle overlapping associated with high doses of adsorbent [45]. Consequently, the exposed surface can be reduced, which leads to a limited number of adsorption sites available for the contaminant [46]. This reduces the adsorption capacity from 18 to 0.6 mg/g for TCS, and from 0.58 to 0.15 mg/g for CAF. Finally, the optimum adsorbent dosages selected through ANOVA statistical analyses (p <0.05) were 8 and 24 g/L for TCS and CAF, respectively.

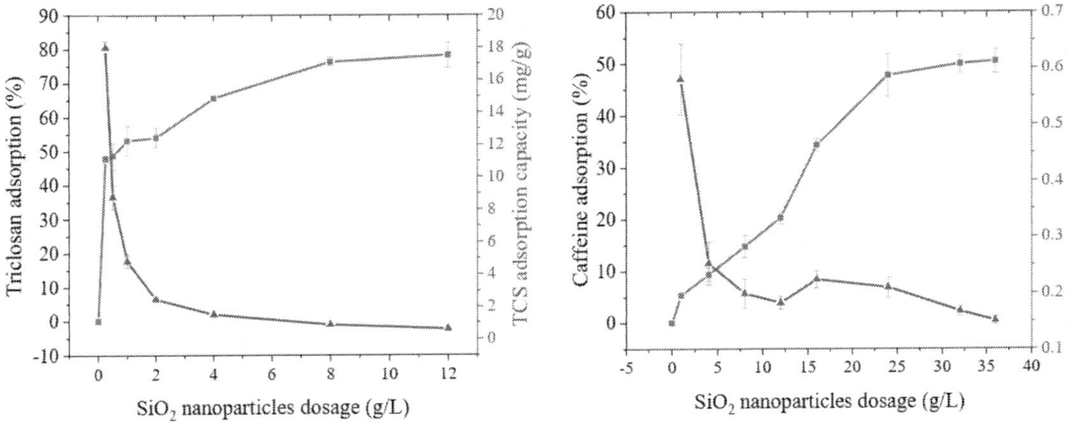

Figure 6. TCS and CAF adsorptive efficiencies as functions of the dose of SiO_2 NPs.

3.2.2. Effect of the contact time. Different time intervals, ranging from 0 to 120 min for CAF, and from 0 to 150 min for TCS, were considered. Figure 7 shows a rapid removal of CAF, going from 38.48 to 53.14 %, during the first 15 minutes of the batch tests. However, increasing the contact time can result in a more complete adsorption equilibrium, observed after 30 min (47.6 to 49.04 %). On the other hand, TCS adsorption shows an increase from 55.04 to 72.57% at 30 min, although the adsorption process achieved equilibrium in 60 min. Initially, the number of vacant sites on the adsorbent surface and of free molecules of the contaminant is high, so adsorption takes place immediately after the contact between the adsorbent and adsorbate [47]. Nevertheless, adsorption becomes less efficient as these sites become occupied, which does not lead to further adsorption [1,48]. Thus, adsorption efficiency decreases to 39.38% and 67.18% at 120 and 150 min for CAF and TCS, respectively. Lastly, the optimum contact time found through ANOVA statistical analyses (p <0.05) were 30 and 60 min for CAF and TCS, respectively, and were used for subsequent experiments.

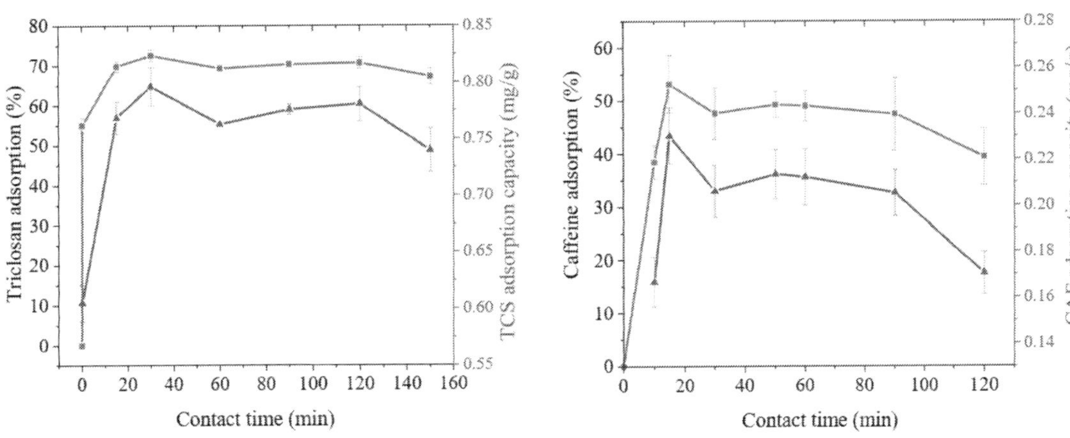

Figure 7. TCS and CAF adsorption as a function of the time of contact.

3.3. Adsorption kinetics

The adsorption dynamics of pollutants can be explained through kinetics studies [22]. The parameters and fitted data for the kinetics models examined are presented in Table 3 and Figure 8. A good fitting to the pseudo-first order model was obtained for CAF adsorption, with $R^2 = 0.92$. This indicates that adsorption occurs immediately in the first few minutes of the process. As time progresses, adsorption gradually increased until no more changes are observed, which shows that the process has reached equilibrium [49]. For the case of TCS, the adsorption on the silica NPs fits better the pseudo-second order kinetics, with $R^2 = 0.99$. In this case, the adsorptive removal slows down near equilibrium, due to the diffusion of contaminant molecules between the walls of the porous medium [50]. According to the kinetics curves, there is a rapid rising at the initial stage (less than 15 min), and then there is a flattening of the curves for a longer adsorption time (Figure 7a).

Table 3. Kinetics parameters for the adsorption of TCS and CAF on silica nanoparticles.

Kinetics models	Model parameters	Triclosan	Caffeine
Experimental data	q_e (mg/g)	0.7731	0.2137
Pseudo-first order	q_e (mg/g)	0.7576	0.2055
	k_1 (1/min)	9.6024	0.2225
	R^2	0.9958	0.9201
Pseudo-second order	q_e (mg/g)	0.7584	0.2055
	k_2 (g/mg min)	31.1607	7.5354
	R^2	0.9959	0.9041
Intra-particle diffusion	kp_1 (mg/ g min$^{-1/2}$)	0.0343	0.0155
	C_1 (mg/g)	0.5955	0.1248
	R^2	0.98184	0.8177
	kp_2 (mg/g min$^{-1/2}$)	0.0109	0.0105
	C_2 (mg/g)	0.6605	0.2929
	R^2	0.7921	0.8151

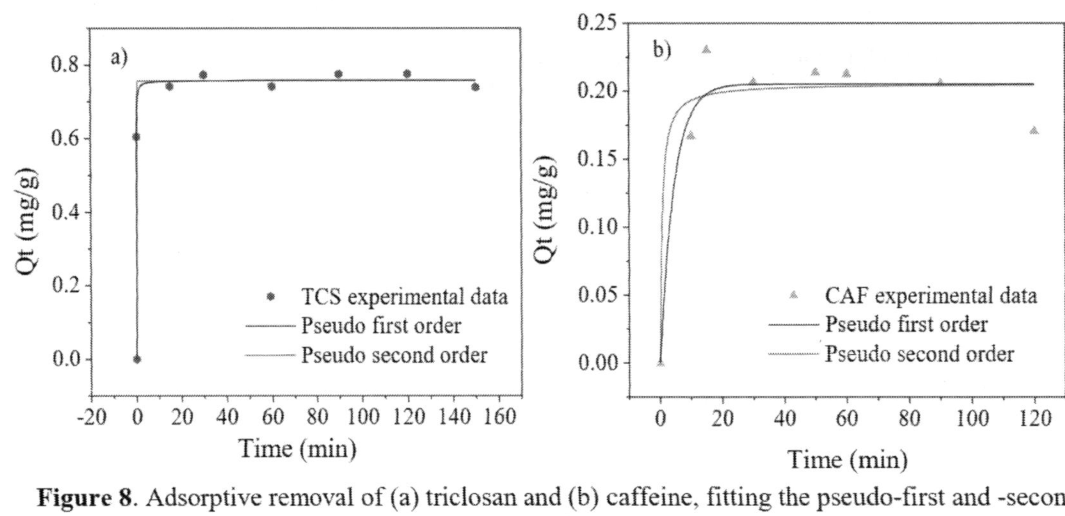

Figure 8. Adsorptive removal of (a) triclosan and (b) caffeine, fitting the pseudo-first and -second order kinetics models.

The intra-particle diffusion model considers the mass transport of a contaminant from the solution to the liquid-solid interface, as well as the diffusion through the porous structure of the adsorbent [9,51]. As shown in Figure 9, the adsorption of CAF and TCS would be governed by multiple processes. This is because the experimental data cannot be described by a single line, and the lines fitted do not pass through the origin either [36]. In this case, when the SiO_2 NPs are dispersed in the synthetic polluted water, the contaminant is immediately transported to the active sites of the adsorbent [52]. Once the outer layers are saturated, the pollutant molecules gradually enter the pores of the silica NPs. This process is known as intra-particle diffusion [53].

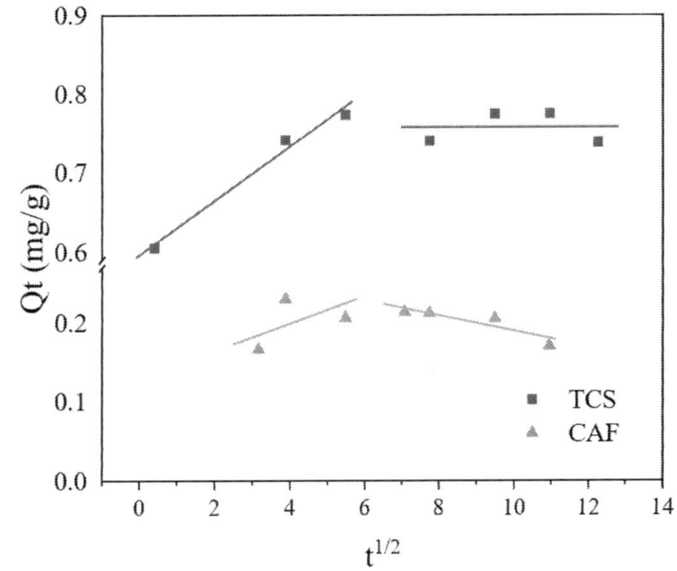

Figure 9. TCS and CAF adsorptive data fitted to the intra-particle diffusion kinetics model.

3.4. Adsorption isotherms

Table 4 and Figure 10 show the coefficients and parameters determined for the fittings of the Langmuir, Freundlich, Sips, and Temkin models to the experimental data collected during the batch tests performed. Results show that the Langmuir isotherm fits well the experimental data of CAF adsorption, with $R^2 = 0.968$. The Langmuir model assumes uniform adsorption on the surface of the adsorbent with a limited number of active sites available and with equal energy for adsorption [36]. The Langmuir q_{max} value was 0.898 mg/g, and K_L was 0.0648. Langmuir's model has a parameter R_L [$R_L = 1/(1 + K_LC_0)$] that indicates the favorability of adsorption. This value is between 0.23 and 0.95, which shows favorable adsorption of caffeine since the factor is between 0 and 1. For the TCS adsorption, Freundlich isotherm presents a good fitting ($R^2 = 0.968$), which describes multilayer adsorption on the heterogeneous surface of silica nanoparticles [49]. The heterogeneity factor n is used to identify the favorability of adsorption, which may be linear (n = 1), favorable (n <1) or unfavorable (n > 1) [53]. According to the results obtained, the adsorption of TCS is favorable (n = 0.82). In addition, the Temkin isotherm model fits well the experimental adsorption data ($R^2 = 0.983$), suggesting that once the adsorption sites are occupied, the process temperature begins to decrease linearly [9]. B_T is the Temkin constant ($B_T = R_T/b$), which takes a positive value of 1.147 and indicates an exothermic process [36]. However, the Sips model is found to best represent the equilibrium data of TCS and CAF. The Sips isotherm considers the competitive effect and the heterogeneity of the adsorbent from the Langmuir and Freundlich models [1]. The high correlations factors R^2 were between 0.977 and 0.997, which show heterogeneous adsorption of CAF and TCS on SiO_2 NPs surface [54]. For CAF adsorption, the R^2 values of the four models decrease in the following order: Sips, Langmuir, Freundlich and Temkin. The adsorption of TCS shows a good fitting in the following order: Sips, Temkin, Freundlich and Langmuir.

Table 4. Parameters calculated from the Langmuir, Freundlich, Sips and Temkin adsorption isotherm models for triclosan and caffeine adsorption.

Isotherm models	Model parameters	Triclosan	Caffeine
Langmuir	q_{max} (mg/g)	5141.7	0.898
	K_L (L/mg)	5.12 x 10^{-5}	0.065
	R^2	0.949	0.968
Freundlich	K_f[(mg/g) $(L/mg)^{1/n}$]	0.178	0.095
	N	0.825	1.857
	R^2	0.968	0.942
Sips	q_m (mg/g)	2.737	0.749
	k_s (L/mg)	0.213	0.096
	M	0.417	0.769
	R^2	0.997	0.977
Temkin	A_T (L/mg)	0.739	1.53
	b (J/ mol)	2160.81	16989
	R^2	0.9826	0.904

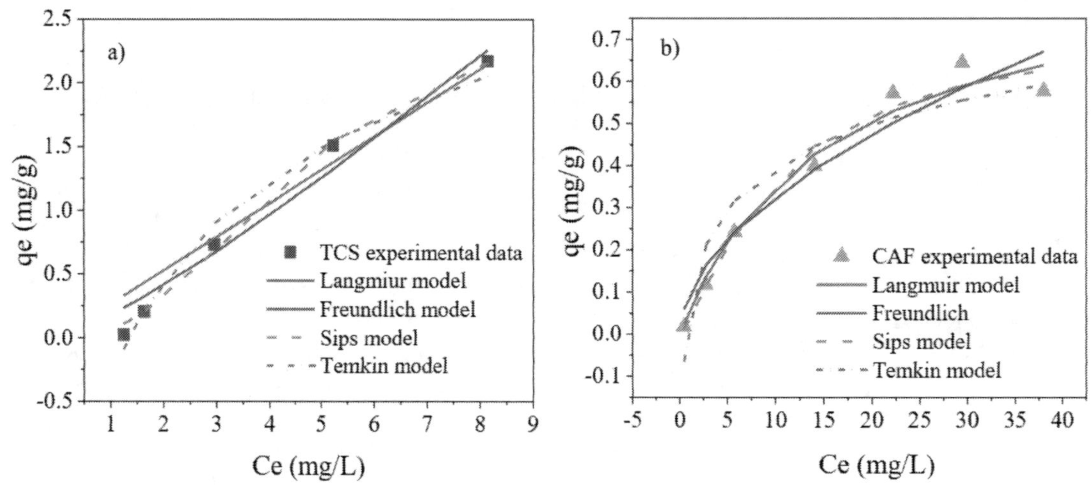

Figure 10. Experimental data and adsorption isotherm models fitted for the removal of (a) triclosan and (b) caffeine.

4. Conclusions

Nowadays, there is an increasing interest in the production of high-added value materials from renewable resources, considering particularly waste materials such as agricultural residues. In this work, nanosilica was synthesized from rice husk following an environmentally friendly approach. This approach consisted in performing an acid leaching treatment of the low-cost natural precursor selected, followed by pyrolysis. Organic acetic and oxalic acids were used instead of the potentially toxic hydrochloric acid traditionally employed in similar syntheses. TEM analyses revealed an average particle size of 5.4, 9.6 and 4.6 nm for the SiO_2-AA, SiO_2-OA and SiO_2-HCl nanoparticles obtained, respectively. SEM images showed strong particle agglomeration and remnants of rice husk residues, as a result of the pyrolysis process. Nevertheless, these aggregates do not seem to influence the specific surface area. BET analyses showed that the silica nanoparticles synthesized show high specific surface areas, between 208 to 223 m^2/g. Furthermore, the materials obtained show an amorphous structure verified by the presence of diffraction halos in SAED pattern and a broad peak around $2\theta \approx 21$ - 23° in the XRD diffraction patterns. The SiO_2-AA nanoparticles synthesized by the acetic acid pretreatment were used as adsorbent for CAF and TCS removal due to their large specific surface area and also its preparation method is relatively simple and less toxic to apply. The batch adsorption studies performed showed that the Sips model fits well the data obtained for both pollutants. The maximum adsorption capacity values of CAF and TCS were 0.75 and 2.74 mg/g, respectively. The optimum dosage and maximum removal efficiencies were 8 g/L and 76%, and 24 g/L and 48%, for TCS and CAF, respectively. The adsorption kinetics of the target pollutants was studied by using three kinetics models. The kinetics data for CAF followed the pseudo-first order kinetics model. Whereas, adsorption of TCS on silica NPs followed a pseudo-second order kinetics. Lastly, the IPD model suggests that adsorption process was governed by a multiple-step mechanism. The promising results obtained open the possibility of evaluating the use of the synthesized nanomaterials from green resources for the adsorptive removal of toxic recalcitrant pollutants.

Acknowledgements
The authors gratefully acknowledge the support provided by the Escuela Politécnica Nacional through the project grant PIGR-19-05. We would also like to thank the Condensed Matter Laboratory of the Department of Physics, and the Department of Extractive Metallurgy of EPN, for their help characterizing the materials used in this work.

References

[1] Bachmann S A L, Calvete T and Féris L A 2021 Caffeine removal from aqueous media by adsorption: An overview of adsorbents evolution and the kinetic, equilibrium and thermodynamic studies *Sci. Total Environ.* **767**

[2] Cusioli L F, Quesada H B, Barbosa de Andrade M, Gomes R G and Bergamasco R 2021 Application of a novel low-cost adsorbent functioned with iron oxide nanoparticles for the removal of triclosan present in contaminated water *Microporous Mesoporous Mater.* **325**

[3] Dvorakova T, Vymazal J, Kozeluh M and Kule L 2018 Occurrence and removal of ibuprofen and its metabolites in full-scale constructed wetlands treating municipal wastewater **120** 1–5

[4] Koc A, Orhon K B, Ogutverici A, Yılmaz L, Oggioni M R, Dilek F B and Yetis U 2014 Is adsorption an artifact in experimentation with Triclosan ? **3994**

[5] Heidler J and Halden R U 2007 Mass balance assessment of triclosan removal during conventional sewage treatment *Chemosphere* **66** 362–9

[6] Song X, Wang X and Bhandari R K 2020 Developmental abnormalities and epigenetic alterations in medaka (Oryzias latipes) embryos induced by triclosan exposure *Chemosphere* **261** 127613

[7] Rigueto C V T, Nazari M T, De Souza C F, Cadore J S, Brião V B and Piccin J S 2020 Alternative techniques for caffeine removal from wastewater: An overview of opportunities and challenges *J. Water Process Eng.* **35**

[8] Moore M T, Greenway S L, Farris J L and Guerra B 2008 Assessing caffeine as an emerging environmental concern using conventional approaches *Arch. Environ. Contam. Toxicol.* **54** 31–5

[9] Rathi B S and Kumar P S 2021 Application of adsorption process for effective removal of emerging contaminants from water and wastewater * *Environ. Pollut.* **280** 116995

[10] Jadhav S A, Garud H B, Patil A H, Patil G D, Patil C R, Dongale T D and Patil P S 2019 Recent advancements in silica nanoparticles based technologies for removal of dyes from water *Colloids Interface Sci. Commun.* **30**

[11] Negrete-Bolagay D, Zamora-Ledezma C, Chuya-Sumba C, De Sousa F B, Whitehead D, Alexis F and Guerrero V H 2021 Persistent organic pollutants: The trade-off between potential risks and sustainable remediation methods *J. Environ. Manage.* **300** 113737

[12] Singh P, Srivastava S and Singh S K 2019 Nanosilica: Recent Progress in Synthesis,

Functionalization, Biocompatibility, and Biomedical Applications *ACS Biomater. Sci. Eng.* **5** 4882–98
[13] Guo Q, Yang G, Huang D, Cao W, Ge L and Li L 2018 Synthesis and characterization of spherical silica nanoparticles by modified Stöber process assisted by slow-hydrolysis catalyst *Colloid Polym. Sci.* **296** 379–84
[14] Zhang Y, Han M-Y and Selvan S T *Properties and Applications 400 5.14.2 Synthesis of Nanoparticles (NPs) Microemulsion 403 5.14.2.1 Synthesis of Metal and Mixed-Metal NPs*
[15] Jeelani P G, Mulay P, Venkat R and Ramalingam C 2020 Multifaceted Application of Silica Nanoparticles. A Review *Silicon* **12** 1337–54
[16] Rajput N 2015 *Methods Of Preparation Of Nanoparticles-A Review* vol 7
[17] Parashar M, Shukla V K and Singh R 2020 Metal oxides nanoparticles via sol–gel method: a review on synthesis, characterization and applications *J. Mater. Sci. Mater. Electron.* **31** 3729–49
[18] Wang W, Martin J C, Fan X, Han A, Luo Z and Sun L 2012 Silica nanoparticles and frameworks from rice husk biomass *ACS Appl. Mater. Interfaces* **4** 977–81
[19] Peerzada J G and Chidambaram R 2021 A Statistical Approach for Biogenic Synthesis of Nano-Silica from Different Agro-Wastes *Silicon* **13** 2089–101
[20] Santana Costa J A and Paranhos C M 2018 Systematic evaluation of amorphous silica production from rice husk ashes *J. Clean. Prod.* **192** 688–97
[21] Pillai P, Dharaskar S, Shah M and Sultania R 2020 Determination of fluoride removal using silica nano adsorbent modified by rice husk from water *Groundw. Sustain. Dev.* **11** 100423
[22] Usgodaarachchi L, Thambiliyagodage C, Wijesekera R and Bakker M G 2021 Synthesis of mesoporous silica nanoparticles derived from rice husk and surface-controlled amine functionalization for efficient adsorption of methylene blue from aqueous solution *Curr. Res. Green Sustain. Chem.* **4** 100116
[23] Gupta A D, Bhadauria V and Singh H 2021 Silica derived from rice husk ash and loaded with iron oxide for As (III) adsorption from water : experimental and modelling studies *Int. J. Environ. Anal. Chem.* **00** 1–24
[24] He J R, Kuo W C, Su C S and Lin H P 2014 Isolation of bio-mesoporous silica from rice husk *J. Chinese Chem. Soc.* **61** 836–40
[25] Wang W, Martin J C, Zhang N, Ma C, Han A and Sun L 2011 Harvesting silica nanoparticles from rice husks *J. Nanoparticle Res.* **13** 6981–90
[26] Karami A, Sabouni R and Ghommem M 2020 Experimental investigation of competitive co-adsorption of naproxen and diclofenac from water by an aluminum-based metal-organic framework *J. Mol. Liq.* **305**
[27] Song J, Zou W, Bian Y, Su F and Han R 2011 Adsorption characteristics of methylene blue by peanut husk in batch and column modes *Desalination* **265** 119–25
[28] Carmona V B, Oliveira R M, Silva W T L, Mattoso L H C and Marconcini J M 2013 Nanosilica from rice husk: Extraction and characterization *Ind. Crops Prod.* **43** 291–6
[29] Zou Y and Yang T 2019 Rice husk, rice husk ash and their applications *Rice Bran and Rice Bran Oil: Chemistry, Processing and Utilization* (Elsevier) pp 207–46
[30] Azat S, Korobeinyk A V., Moustakas K and Inglezakis V J 2019 Sustainable production of pure silica from rice husk waste in Kazakhstan *J. Clean. Prod.* **217** 352–9
[31] Gottardi V, Guglielmi M, Bertoluzza A, Fagnano C and Morelli M A 1984 *Further Investigations On Raman Spectra Of Silica Gel Evolving Toward Glass* vol 63
[32] Adebisi J A, Agunsoye J O, Bello S A, Haris M, Ramakokovhu M M, Daramola M O and Hassan S B 2020 Green production of silica nanoparticles from maize stalk *Part. Sci. Technol.* **38** 667–75
[33] McMnr quL F and Rruurr R L 1986 *Hydroxyl sites in SiO, glass: A note on infrared and Raman spectra* vol 71
[34] Rafigh S M and Heydarinasab A 2017 Mesoporous Chitosan-SiO2 Nanoparticles: Synthesis,

Characterization, and CO2 Adsorption Capacity *ACS Sustain. Chem. Eng.* **5** 10379–86

[35] Sankar S, Sharma S K, Kaur N, Lee B, Kim D Y, Lee S and Jung H 2016 Biogenerated silica nanoparticles synthesized from sticky, red, and brown rice husk ashes by a chemical method *Ceram. Int.* **42** 4875–85

[36] Karim A H, Jalil A A, Triwahyono S, Sidik S M, Kamarudin N H N, Jusoh R, Jusoh N W C and Hameed B H 2012 Amino modified mesostructured silica nanoparticles for efficient adsorption of methylene blue *J. Colloid Interface Sci.* **386** 307–14

[37] Bakar R A, Yahya R and Gan S N 2016 Production of High Purity Amorphous Silica from Rice Husk *Procedia Chem.* **19** 189–95

[38] Azat S, Korobeinyk A V., Moustakas K and Inglezakis V J 2019 Sustainable production of pure silica from rice husk waste in Kazakhstan *J. Clean. Prod.* **217** 352–9

[39] Ahmadi A, Sedaghat T, Azadi R and Motamedi H 2020 Magnetic Mesoporous Silica Nanocomposite Functionalized with Palladium Schiff Base Complex: Synthesis, Characterization, Catalytic Efficacy in the Suzuki–Miyaura Reaction and α-Amylase Immobilization *Catal. Letters* **150** 112–26

[40] Wei T Y, Kuo C Y, Hsu Y J, Lu S Y and Chang Y C 2008 Tin oxide nanocrystals embedded in silica aerogel: Photoluminescence and photocatalysis *Microporous Mesoporous Mater.* **112** 580–8

[41] Beltrame K K, Cazetta A L, de Souza P S C, Spessato L, Silva T L and Almeida V C 2018 Adsorption of caffeine on mesoporous activated carbon fibers prepared from pineapple plant leaves *Ecotoxicol. Environ. Saf.* **147** 64–71

[42] Almeida S R, Elicker C, Vieira B M, Cabral T H, Silva A F, Sanches Filho P J, Raubach C W, Hartwig C A, Mesko M F, Moreira M L and Cava S 2019 Black SiO 2 nanoparticles obtained by pyrolysis of rice husk *Dye. Pigment.* **164** 272–8

[43] Alyosef H A, Eilert A, Welscher J, Ibrahim S S, Denecke R, Schwieger W and Enke D 2013 Characterization of biogenic silica generated by thermo chemical treatment of rice husk *Part. Sci. Technol.* **31** 524–32

[44] Sheikhmohammadi A, Safari M, Alinejad A, Esrafili A, Nourmoradi H and Asgari E 2019 The synthesis and application of the Fe3O4@SiO2 nanoparticles functionalized with 3-aminopropyltriethoxysilane as an efficient sorbent for the adsorption of ethylparaben from wastewater: Synthesis, kinetic, thermodynamic and equilibrium studies *J. Environ. Chem. Eng.* **7**

[45] Kaur H, Bansiwal A, Hippargi G and Pophali G R 2018 Effect of hydrophobicity of pharmaceuticals and personal care products for adsorption on activated carbon: Adsorption isotherms, kinetics and mechanism *Environ. Sci. Pollut. Res.* **25** 20473–85

[46] Anastopoulos I, Pashalidis I, Orfanos A G, Manariotis I D, Tatarchuk T, Sellaoui L, Bonilla-Petriciolet A, Mittal A and Núñez-Delgado A 2020 Removal of caffeine, nicotine and amoxicillin from (waste)waters by various adsorbents. A review *J. Environ. Manage.* **261**

[47] Danish M 2020 Application of date stone activated carbon for the removal of caffeine molecules from water *Materials Today: Proceedings* vol 31 (Elsevier Ltd) pp 18–22

[48] Farmany A, Mortazavi S S and Mahdavi H 2016 Ultrasond-assisted synthesis of Fe3O4/SiO2 core/shell with enhanced adsorption capacity for diazinon removal *J. Magn. Magn. Mater.* **416** 75–80

[49] Oliveira M F, Silva M G C and Vieira M G A 2019 Applied Clay Science Equilibrium and kinetic studies of ca ff eine adsorption from aqueous solutions on thermally modi fi ed Verde-lodo bentonite *Appl. Clay Sci.* **168** 366–73

[50] Simonin J 2016 On the comparison of pseudo-first order and pseudo-second order rate laws in the modeling of adsorption kinetics *Chem. Eng. J.* **300** 254–63

[51] Czech B, Shirvanimoghaddam K, Trojanowska E and naebe M 2020 Sorption of pharmaceuticals and personal care products (PPCPs) onto a sustainable cotton based adsorbent *Sustain. Chem. Pharm.* **18**

[52] Önal Y, Akmil-Başar C and Sarici-Özdemir Ç 2007 Elucidation of the naproxen sodium adsorption onto activated carbon prepared from waste apricot: Kinetic, equilibrium and thermodynamic characterization *J. Hazard. Mater.* **148** 727–34
[53] Tran H N, You S, Hosseini-bandegharaei A and Chao H 2017 Mistakes and inconsistencies regarding adsorption of contaminants from aqueous solutions: A critical review *Water Res.*
[54] Ahmed M J and Dhedan S K 2012 Fluid Phase Equilibria Equilibrium isotherms and kinetics modeling of methylene blue adsorption on agricultural wastes-based activated carbons *Fluid Phase Equilib.* **317** 9–14

Journal of Physics: Conference Series **2238** (2022) 012006 doi:10.1088/1742-6596/2238/1/012006

Photocatalytic activity of Fe/Ti-based compounds obtained from ferro-titaniferous mineral sands via a simple soft chemistry route

Salomé Galeas[1,2], Carla S. Valdivieso-Ramírez[1], Patricia I. Pontón[1], Vincent Goetz[3] and Victor H. Guerrero[1,*]

[1] Department of Materials, Escuela Politécnica Nacional, Quito 170525, Ecuador
[2] University of Perpignan Via Domitia, 52 Paul Alduy, 66100 Perpignan, France
[3] PROMES CNRS, UPR 8521, Rambla de la Thermodynamique, 66100 Perpignan, France

*Corresponding author: victor.guerrero@epn.edu.ec

Abstract. Fe/Ti-based compounds were synthesized from ferro-titaniferous mineral sands using aqueous oxalic acid under subcritical water (sCW) conditions (135°C/50 bar/4 h) and their photocatalytic activity was evaluated against methylene blue or caffeine under UV-A and visible radiation. The X-ray diffraction and Raman spectroscopy analyses revealed that the as-synthetized compounds were mainly comprised of ferrous oxalate, followed by titanium dioxide. This proved the complete transformation of the mineral precursor within 4 h using oxalic acid/sCW as the reaction medium. The photocatalytic studies showed that methylene blue and caffeine were photodegraded under UV-A and visible radiation by the as-synthesized Fe/Ti-based compounds. Methylene blue removal was up to 92.8 and 97.4% after 6 hours under UV-A or visible irradiation, respectively. Caffeine removal, however, reached 69.3 and 59.6% after 16 h of exposure to UV-A and visible light, respectively. The as-synthesized compounds can be potentially used as a ferrous-based catalyst in heterogeneous photocatalysis for decontamination of water systems from organic pollutants.

1. Introduction

Water pollution has become a serious concern worldwide. This pollution is caused by industrial effluents, hospitals and domestic sewage, and threatens water quality and reduces the availability of clean water globally, demanding the development of new materials and technologies for its treatment and decontamination [1]. During the last decades, several processes for removal and degradation of organic pollutants present in water have been widely studied [2–5]. However, conventional treatments as flotation, reverse osmosis, adsorption, ultrafiltration, or biological treatment, are not always efficient and could be energy-consuming [4,6,7].

Advanced oxidation processes (AOPs) have emerged as an alternative for water treatment, as their aim is the complete degradation of organic pollutants into harmless compounds due to the in-situ generation of highly reactive oxygen species (ROS) such as sulfate and hydroxyl radicals [8]. Among these AOPs, photocatalysis is a promising technique, which consists in the generation of hydroxyl radicals by a photocatalytic material under irradiation, leading to degradation of trace organic chemicals [9]. These photocatalytic materials are even more attractive if they can be obtained from widely available and low-cost precursors by using energy efficient and environmentally friendly methods that do not require toxic and expensive solvents [10–13]. Iron-based materials, for example, are widely used in heterogeneous photocatalysis because of their effectiveness, abundance, and low cost [14]. Among iron-based materials, iron oxalate can be obtained by simple reaction of oxalic acid with an iron source, which photocatalytic activity has been reported in literature. In addition, recovery of ferrous oxalate from wastewater streams of washing of heavy metal contaminated soils has been also explored [15] as well as its photocatalytic activity studied on methylene blue (MB) [15]. As such, degradation of MB (20 ppm solution) up to 50 and 90% was reported within 1 and 3 h, respectively, when 0.5 g/L of recovered ferrous oxalate was used as catalyst under simulated sun light [15]. Also, Hu et al. [16] used ferrous oxalate to degrade MB but in Fenton and photo-Fenton processes. In their case, a combination of 0.5 g/L of catalyst and hydrogen peroxide (30 mg/L) led to 94% and 97.2% of MB (10 ppm solution) degradation within 10 and 15 min, respectively [16]. Hence, the percent degradation was improved under a xenon lamp irradiation (500 W), reaching over 95% within 3 min, and 98.4% in 10 min [16]. The photocatalytic activity of ferrous oxalate has been also explored for degradation of other cationic and anionic dyes like rhodamine B [17] and indigo red [18], respectively. In addition, ferrous oxalate has proven to be effective in the oxidation of the herbicide 2,4-dichlorophenoxyacetic acid [19]. In this context, MB and caffeine were selected as the model pollutants for this study.

MB dye can be used as a suitable pollutant to assess the photocatalytic activity of iron-based photocatalysts and to get an insight on photodegradation of similar chemical compounds that are resistant to biodegradation and are widely present in the large volume effluents of the textile industry that threaten the ecosystem and human health [20]. Likewise, caffeine was selected since it is a common alkaloid, highly soluble in water, extensively found in water bodies, which can negatively impact aquatic life [21]. Currently, studies focused on either MB or caffeine degradation by ferrous oxalate-based catalysts are limited. Therefore, the aim of this work was to first synthesize Fe/Ti-based compounds from ferro-titaniferous mineral sands using aqueous oxalic acid under subcritical water (sCW) conditions (135°C/50 bar/4 h) as an alternative eco-friendly process to conventional mineral acid leaching and further evaluate whether the as-synthetized compounds exert a photocatalytic activity against methylene blue or caffeine under UV-A and visible radiation.

2. Materials and methods

2.1. Materials

Ecuadorian ferro-titaniferous sands from the "El Ostional" beach were used as mineral precursor. Such ferro-titaniferous mineral was reported as an ilmenite hematite solid solution with estimated chemical composition of $0.6FeTiO_3 \cdot 0.4Fe_2O_3$ [22]. Oxalic acid dihydrate was supplied by Merck, caffeine was purchased from Beantown Chemicals and methylene blue from Panreac. All chemicals were analytical grade and used as received. Water from the Milli-Q system (Millipore, Billerica, MA, USA) and N_2 gas (99.9% purity) from Linde (Quito, Ecuador) were used for the subcritical water synthesis.

2.2. Synthesis of the catalysts

The synthesis of the iron-based catalysts from the ferro-titaniferous sands was carried out in a batch stirred high pressure reactor (Berghof BR-500, Germany) following the methodology described by Valdivieso-Ramírez et al. [23], with some modifications. Briefly, 300 g of 1 M oxalic acid solution and 1.75 g of ferro-titaniferous sands were loaded into the reactor. Once the system was assembled and N_2 pressurized, the reaction was set at 135°C, 50 bar and 4 h. The as-synthesized product stream was oven-dried at 60°C for 12 h and stored for characterization and further utilization in the photocatalytic activity test. The sCW synthesis of the catalyst is summarized in Figure 1.

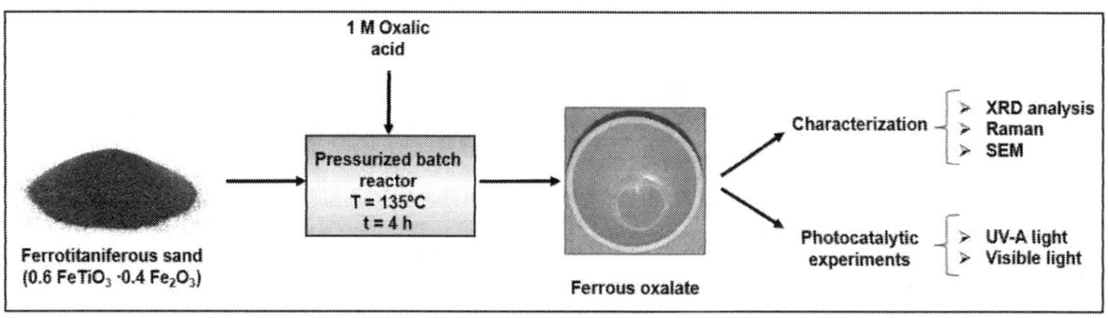

Figure 1. Scheme for the synthesis of iron-based photocatalyst from ferro-titaniferous sands with 1 M oxalic acid at 135° and 50 bar.

2.3. Characterization

X-ray diffraction (XRD) analysis was performed on the as-synthesized product using a Bruker D2 Phaser X-ray diffractometer with CuKα radiation (λ = 1.54184 Å), step size 0.02°, with LYNXEYE XE-T detector (1D-mode). Also, the as-synthesized product was analyzed by Raman spectroscopy using a LabRAM HR Evolution confocal microscopic Raman spectrometer (HORIBA Scientific, France) with a 633 nm laser irradiation set at 12.5 mW of power. Scanning electron microscopy (SEM) images of the as-synthetized product were recorded in a PSEM eXpress (ASPEX, USA) operated in secondary electron mode.

2.4. Photocatalytic activity test

The photocatalytic activity of the as-synthetized iron-based catalyst on methylene and caffeine was evaluated under UV-A and visible radiation according to the method depicted in Figure 2. Such method consisted of an initial dispersion of the aqueous pollutant with the solid catalyst by means of an ultrasonic probe (10 kHz, t=1 min), orbital stirring and subsequent analysis of the filtered solutions by UV/VIS spectrometry (HORIBA, Duetta) at 664 and 272 nm for methylene blue and caffeine, respectively.

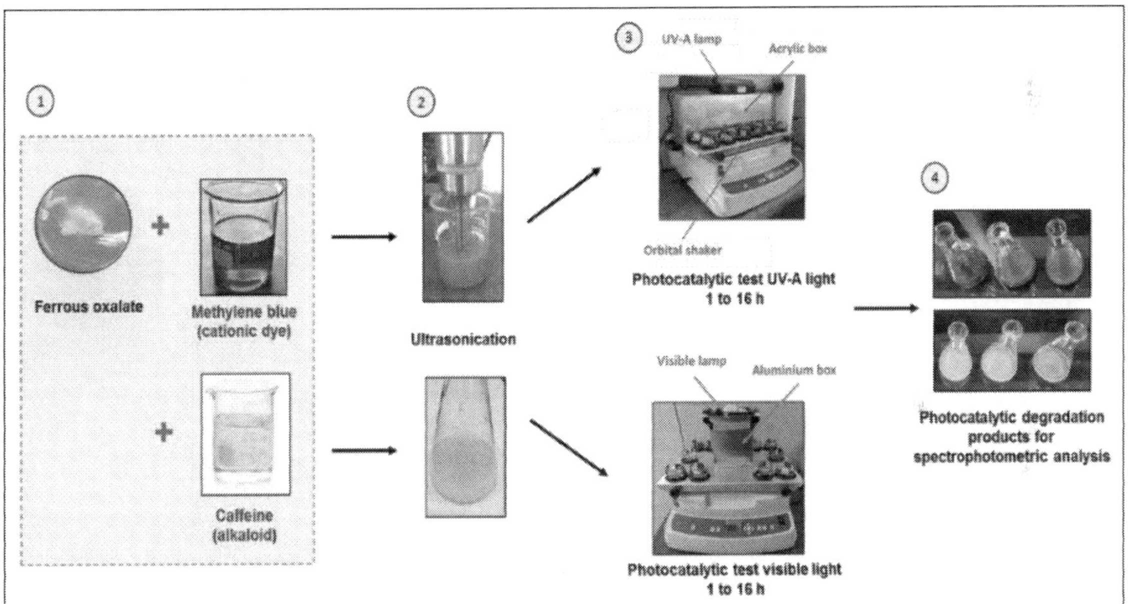

Figure 2. Scheme followed for the photocatalytic tests.

For the photocatalysis test, solutions of methylene blue and caffeine of 10 ppm were used. Also, a working pH of 3.1 was selected based on the reported conditions for Fenton and photocatalytic processes [24]. The catalyst load was 1.5 g/L and the reaction times 1, 3, 6 and 16 h. The photocatalysis was carried out in a chamber equipped with a UV-A or a visible irradiation lamp placed on an orbital shaker set at 150 rpm (TECNAL, model TE-1400). UV-A radiation was supplied by a 6 W UV lamp with light emission at 365 nm (UVP, model UVGL-55), whereas visible light was provided by an 18 W LED lamp (Sylvania, LED panel SQ 18W DL). The experimental conditions used for the photocatalysis of the model pollutants by the as-synthetized iron-based catalyst are detailed in Table 1.

Table 1. Experimental conditions used for the removal of MB and caffeine from aqueous solutions by the as-synthetized iron-based catalyst.

Pollutant	Catalyst	Light source	Reaction times (h)	Experiment
MB 10 ppm	$FeC_2O_4 \cdot 2H_2O$ (135°C-4h) 1.5 g/L	UV-A	1, 3, 6, 16	Photocatalysis
		Visible	1, 3, 6, 16	Photocatalysis
		Darkness	1, 3, 6, 16	Adsorption
	None	UV-A	16	Photolysis
		Visible	16	Photolysis
Caffeine 10 ppm	$FeC_2O_4 \cdot 2H_2O$ (135°C-4h) 1.5 g/L	UV-A	1, 3, 6, 16	Photocatalysis
		Visible	1, 3, 6, 16	Photocatalysis
		Darkness	1, 3, 6, 16	Adsorption
	None	UV-A	16	Photolysis
		Visible	16	Photolysis

3. Results

3.1. Characterization of the as-synthetized catalyst

The X-ray diffractogram of the as-synthesized catalyst is shown in Figure 3. Peaks at diffraction angles (2θ) correspond to ferrous oxalate hydrate $FeC_2O_4 \cdot 2H_2O$ in two crystalline phases, both with the same monoclinic unit cell, but with different values of the lattice parameter (a). Peaks observed at 18.7°, 19.1°, 23.1°, 25.0°, 28.3°, 29.9°, 32.5°, 34.0°, 34.5°, 37.7°, 38.4°, 40.8°, 43.2°, 44.7°, 45.1°, 47.3°, 48.3°, 49.5°, 50.4°, 51.2°, and 57.2°, are in accordance with PDF # 23-0293, where a = 9.84 Å. The diffraction peaks appearing at 35.5°, 40.1°, 42.8°, 46.0°, 46.8°, 53.1°, 56.4°, and 58.1° correspond to PDF # 72-1305, where a = 12.06 Å.

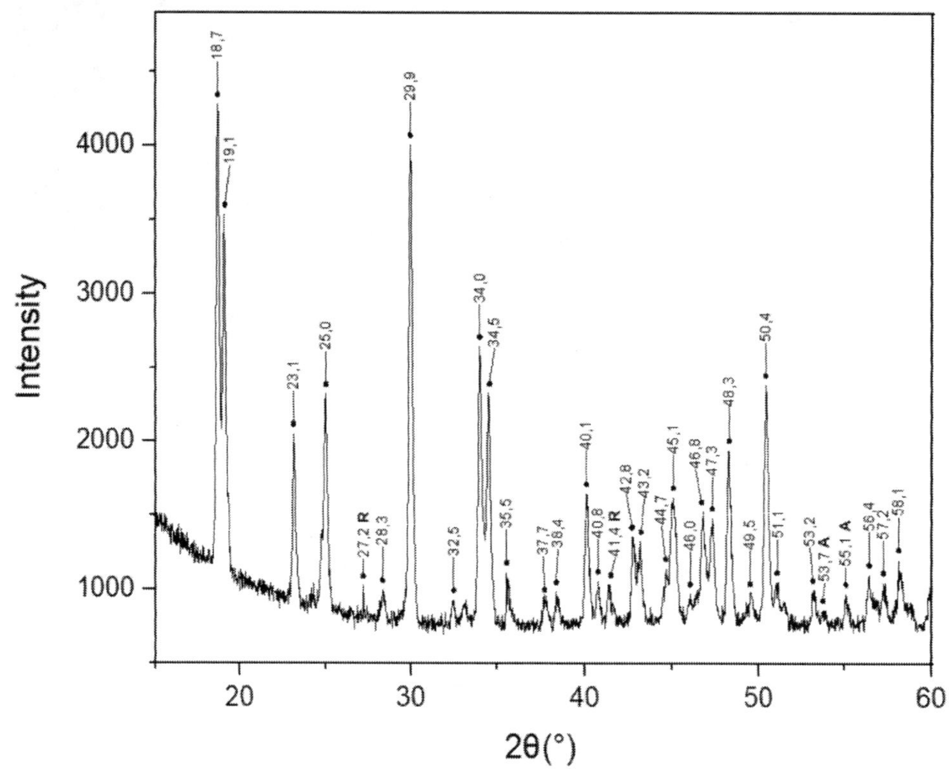

Figure 3. X-ray diffractogram of the as-synthetized product from ferro-titaniferous sands with 1 M oxalic acid at 135°C, 50 bar and 4 h.

The XRD pattern also showed additional peaks which are associated to the presence of titanium dioxide, because Ti is also present in the ferro-titaniferous sand precursor. Diffraction peaks at 2θ values of 27.2° and 41.4°, correspond to rutile phase (PDF 21-1276) and the peaks at 53.7° and 55.1° are ascribed to anatase phase (PDF 21-1272). Since the intensity of these peaks is lower than that of ferrous oxalate, the TiO_2 phases are present in a smaller fraction.

The Raman spectrum of the as-synthetized catalyst, depicted in Figure 4, presents the characteristic bands of ferrous oxalate. The bands observed at 1710 and 1470 cm^{-1} may be assigned to the ν (C=O) stretching mode. Another band at the low wavenumber side of the symmetric stretching mode appeared at 1435 cm^{-1}. The band at 915 cm^{-1} could be attributed to the ν (C–C) stretching mode and the low intensity band at 858 cm^{-1} is assigned to the O-C=O bending mode. The band at 585 cm^{-1} matches with water librational mode and the band at 520 cm^{-1} may be attributed to the symmetric O-C=O bending mode (δ-ring). Bands at 288, 246, and 206 cm^{-1} may be assigned to Fe-O stretching and bending modes [25].

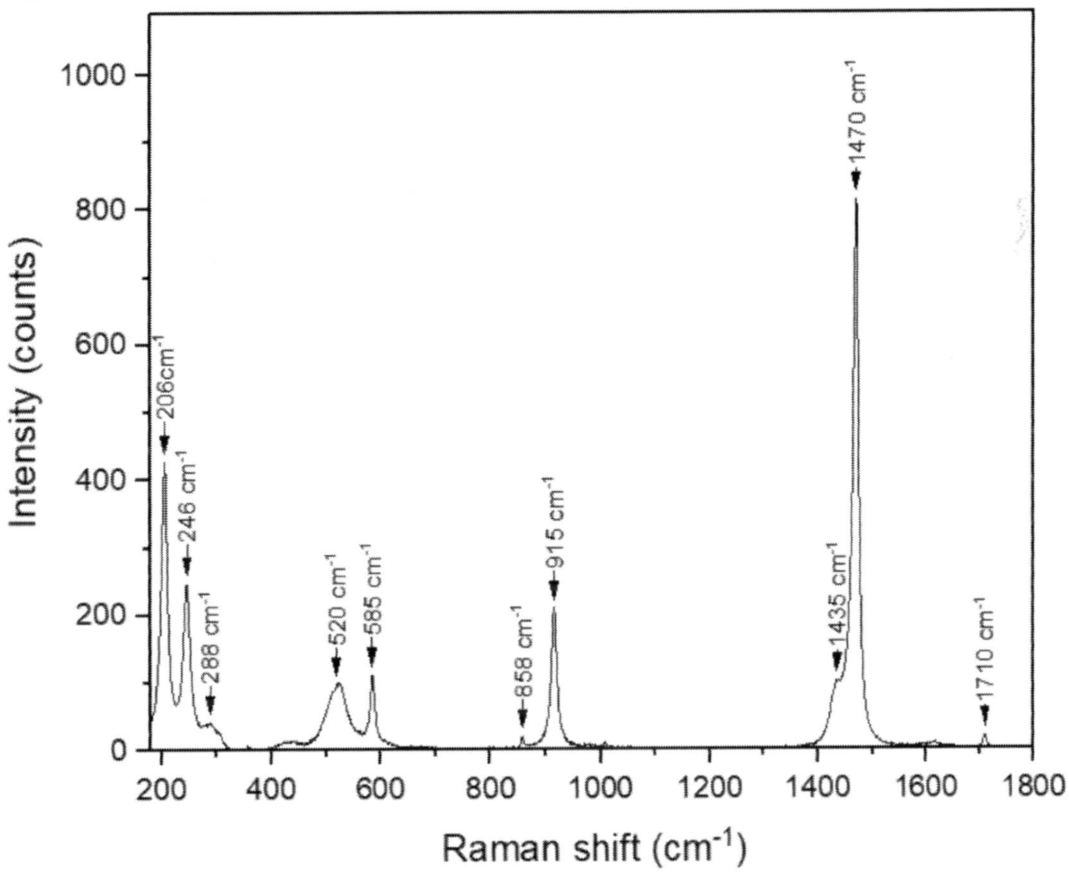

Figure 4. Raman spectra of the as-synthetized product from ferro-titaniferous sands with 1 M oxalic acid at 135°C, 50 bar and 4 h.

In addition, the morphology of the dry as-synthetized product powder was observed by SEM. Figure 5 shows that the as-synthetized product is of heterogenous size and morphology as indicated by the smaller structures with sharp pyramidal shape as well as the presence of submicronic structures with a rhomboidal crystal trend.

The characterization of the as-synthetized product demonstrated that it is mostly composed of ferrous oxalate, which could be potentially used for photocatalysis. Therefore, the photocatalytic activity of the FeC_2O_4-based product was evaluated next.

Figure 5. SEM image of the as-synthetized product from ferro-titaniferous sands with 1 M oxalic acid at 135°C, 50 bar and 4 h.

3.2. Photocatalytic activity

3.2.1. Methylene blue removal. Prior to photocatalytic activity tests under UV-A and visible light (Table 1), an adsorption test at 1, 3, 6 and 16 h of contact time was carried out using a methylene blue pristine solution. According to the UV-VIS spectrograph showed in Figure 6, the absorbance of the pristine MB solution got reduced with time, where major removal of MB (27.6%) occurred within 1 h and after 16 h only an additional 2.3 % was removed. However, the highest percent removal was observed at 6 h (37.9%), suggesting that a desorption phenomenon took place within the next 10 h that led to a final percent removal of 29.9%. In addition, no shift of the band of maximum absorption at 664 nm was apparent with time, indicating that adsorption could be the main catalytic mechanism associated with MB removal by the synthesized FeC_2O_4-based catalyst in the absence of external radiation.

Figure 7 shows the UV-VIS spectra of MB photocatalysis by the FeC_2O_4-based synthetized catalyst. According to Figures 7.a and 7.b, photodegradation of MB indeed occurred at either UVA or visible light conditions as an hypsochromic shift from 664 to 628 nm was apparent while the absorbance values got reduced. Interestingly, photooxidation of MB by a TiO_2/UV catalytic system has reported also a hypsochromic shift due to partial consecutive N-demethylation of MB to blue colored compounds such as Azure B, A and C that absorb light at lower wavelengths of 655-648, 634-620 and 612-608 nm, respectively [26]. Although the catalytic system used in this work (FeC_2O_4/UV-light), differs from the one mentioned above, generation of hydroxyl radicals and further oxidation of MB into blue colored analogous compounds can be also expected as FeC_2O_4 has shown to be a highly photosensitive complex that can assist the generation of hydroxyl radicals in the presence of UV-light [27]. As such, the spectra in Figure 7 show a wavelength shift with time from 664 to 628 nm suggesting that the formation of partially demethylated compounds Azure B and Azure A occurred. Nonetheless, further HPLC analysis are recommended for identification and quantification of the above-mentioned compounds.

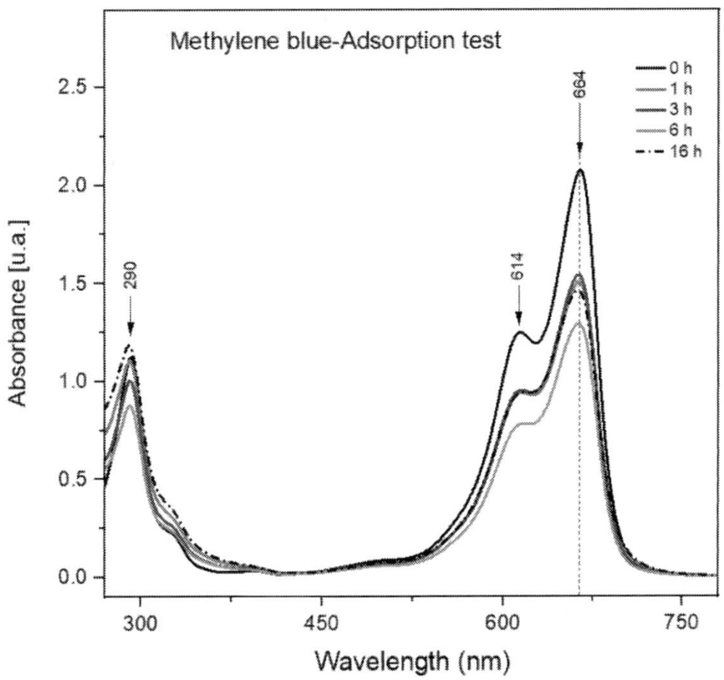

Figure 6. UV-VIS spectra of methylene blue adsorption by the FeC_2O_4-based synthetized catalyst.

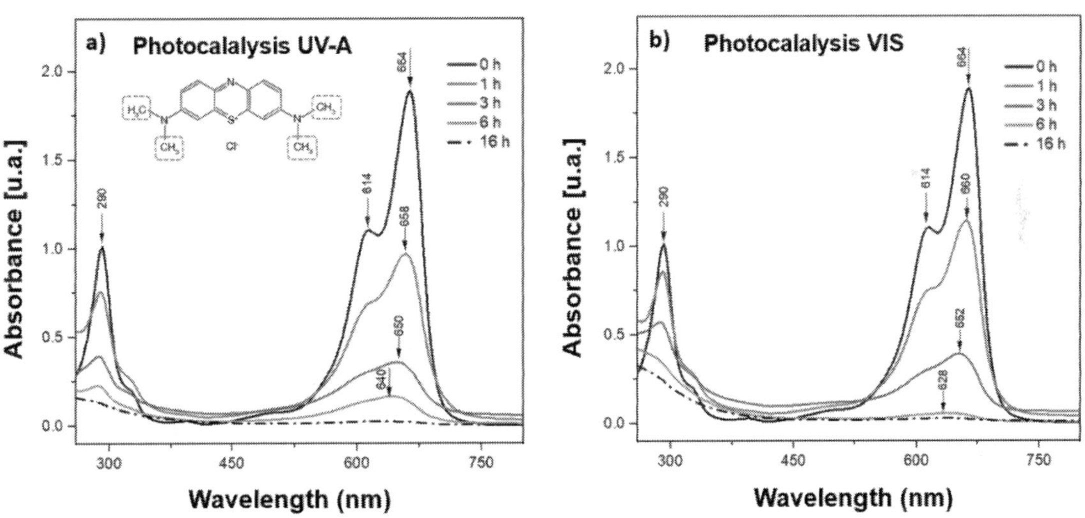

Figure 7. UV-VIS spectra of methylene blue photocatalysis under a) UV-A light and b) visible light by the FeC_2O_4-based synthetized catalyst.

The percent removal of methylene blue under UV-A and visible light radiation is displayed in Figure 8a. According to Figure 8a, the type of radiation source did not influence considerably the removal of methylene blue as the rate of removal was similar for photocatalysis at either UV-A or visible light conditions. As such, percent removal after 6 h reached 93 and 98 % for UV-A and visible light exposure, respectively., and after 16 h, the removal was almost complete. The removal of methylene blue was associated with a gradual change of color intensity, as seen in Figure 8b. In addition, methylene blue removal could be due to adsorption and photodegradation effects as the rate of removal (steepest slope of the curve removal % vs. time) was the highest within the first hour. Conversely, experiments in the absence of the catalyst were done to determine the stability of pristine methylene blue solution upon exposure to the selected radiation. The percent photolysis was 3.7% for both light sources. Based on the above-mentioned results, the as-synthetized FeC_2O_4-based compound exhibited photocatalytic activity under UV-A and visible irradiation, which is very desirable in photocatalyst to effectively utilize the solar energy.

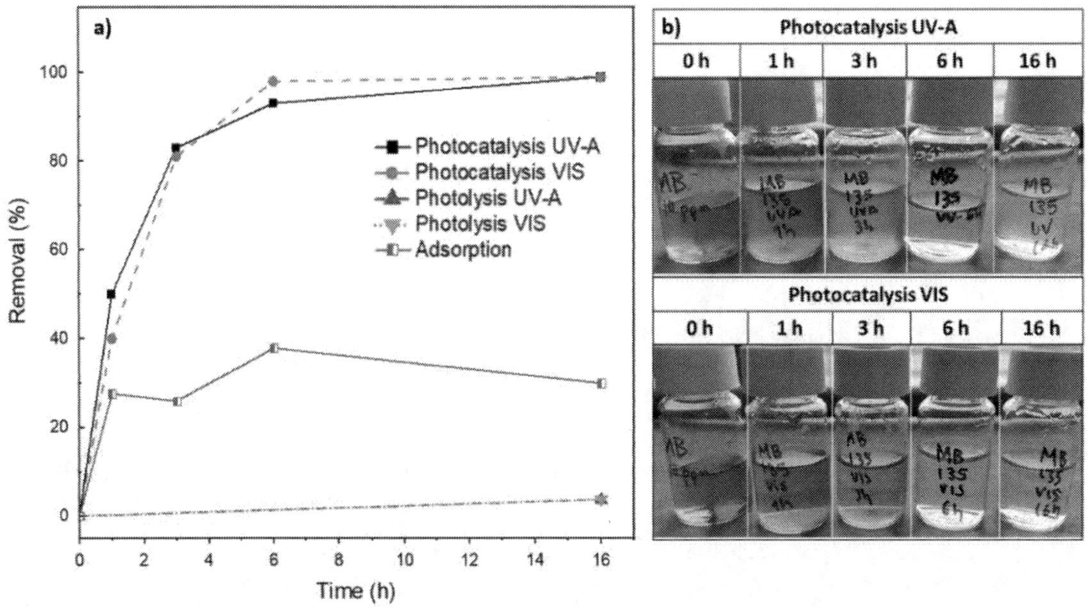

Figure 8. (a) Methylene blue percent removal by absorption, photolysis and photocatalysis via the FeC_2O_4-based synthetized catalyst and b) Methylene blue solutions obtained after photocatalysis under UV-A and visible light irradiation.

3.2.2. Caffeine removal. To evaluate caffeine behavior when in contact with the FeC_2O_4-based synthetized catalyst, an adsorption test at conditions detailed in Table 1 was carried out on caffeine pristine solution. The resulting UV-VIS spectra (Figure 9), showed increased absorbance values with time compared to that of the pristine caffeine solution, suggesting the formation of other compounds possible due to caffeine-iron catalyst interactions. Interestingly, formation of metal-caffeine coordination complexes has been reported for Ca, Mg, Fe, Zn, etc. [28,29]. As such, a binding constant of 59 has been attributed to the Fe^{+2}-caffeine coordination complex [28]. In addition, the absorption of electromagnetic radiation of the iron–caffeine coordination complex has been reported near to the maximum absorption band of pure caffeine at 272 nm [30]. The formation of such coordination complex can resemble to an acid-base reaction where caffeine (Lewis base ligand) donates a pair of electrons to the metal ion Fe^{+2} (Lewis acid) to form the metal-ligand bond [31]. Thus, the iron-caffeine complex can also be considered as a charge-transfer complex, where absorption of radiation is due to an electronic transition from the donor to the corresponding orbital of the acceptor so that the associated molar absorptivity (ε) is high compared to that of molecules that exhibit π, σ and n electron transitions [31]. As such, the increased absorbance values at 272 nm observed in Figure 9 could be associated to the increased light absorption of the newly formed iron-caffeine complex, which molar absorptivity has been reported larger ($>1x10^4$ $M^{-1}cm^{-1}$) than that of pristine caffeine ($>9.74x10^3$ $M^{-1}cm^{-1}$) [32]. Also, as the absorbance is directly proportional to molar absorptivity (ε), the concentration of the absorbing species and the path length (Beer's Law), the increased ε value of the iron-caffeine complex will indeed lead to increased absorbance. Therefore, it can be inferred that caffeine reacts with the iron present in the FeC_2O_4-based catalyst and forms a soluble iron-caffeine complex, which concentration increases with time.

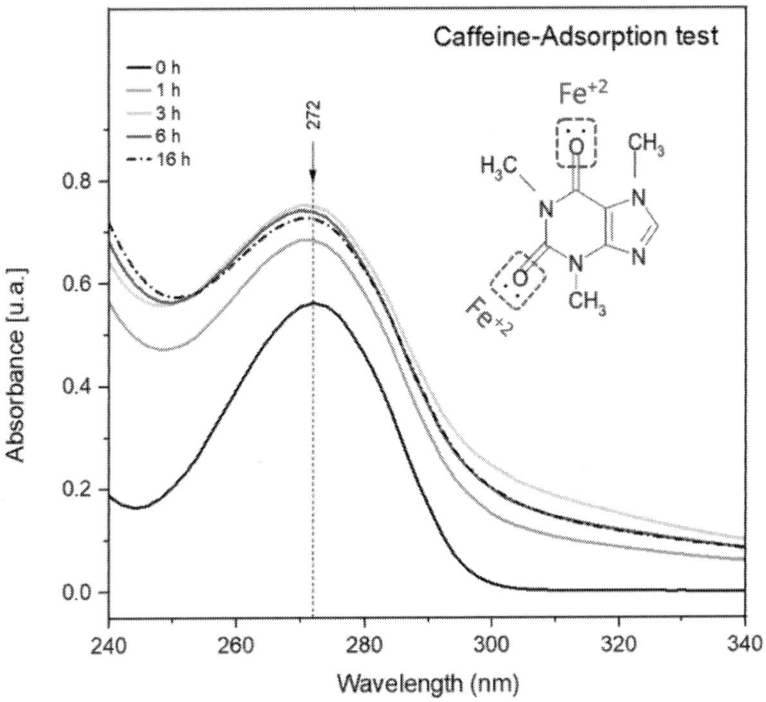

Figure 9. UV-VIS spectra of caffeine adsorption test by the FeC_2O_4-based synthetized catalyst.

Figure 10 shows the UV-VIS spectra of caffeine photocatalysis by the FeC_2O_4-based synthetized catalyst under UV-A or visible light. According to Figures 10.a and 10.b, photodegradation of caffeine indeed occurred as the absorbance values at the maximum absorption wavelength at 272 nm got considerably reduced with time. Also, band broadening was apparent as the time approached to 16 h, possibly due to the formation of intermediate compounds. Demethylation and hydroxylation of caffeine have been reported due to in-situ generation of hydroxyl radicals under Fenton conditions leading to 1,3-dimethyl uric acid and 6-amino-5-(N-formyl methyl amino)-1,3-dimethyl-uracil [33]. Also, the formation of intermediates like 1, 3-dimethyl-2, 4, 5-trioxoimidazolidinetrione and other compounds with mass-to-charge ratio (m/z) of 181.0738, 211.0906 and 227.0834 have been identified by electrospray ionization mass spectrometry (ESI-MS) in the presence of Fe^{+2}/persulfate/UV-light [34], but complete caffeine mineralization (CO_2, H_2O, and NH_3) via the reported photocatalytic systems has to been achieved. As the photocatalytic system used in the present study (FeC_2O_4-TiO_2/UV-A or VIS) resembles to those mentioned above, in the sense of its iron and titanium dioxide content, it could be inferred that the main reactive specie generated during the photocatalysis was the hydroxyl radical; therefore, photooxidation of caffeine could have occurred. In addition, the band broadening exhibited in the UV-VIS spectra obtained from the caffeine solutions after photocatalysis (Figure 10) could be attributed to the presence of intermediates that absorb light at a wide range of wavelengths like the ones similar to uric acid and uracil that exhibit maximum absorption at 290 cm^{-1} [35] and 260 and 300 cm^{-1} [36], respectively.

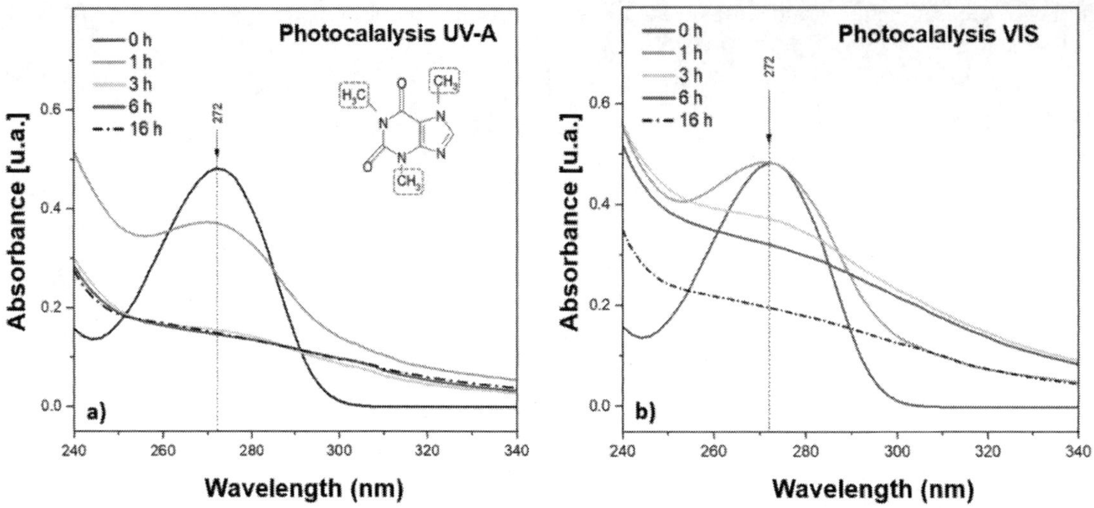

Figure 10. UV-VIS spectra of the solutions after caffeine photocatalysis under a) UV-A light, and b) visible light irradiation by the FeC_2O_4-based synthetized catalyst.

The percent removal of caffeine under UV-A and visible light radiation is displayed in Figure 11. According to Figure 11, the type of radiation source influenced considerably the removal of caffeine as the rate of removal for photocatalysis under UV-A was faster than that of the visible light. As such, percent removal after 6 h reached 70 and 34 % for UV-A and visible light exposure, respectively., and after 16 h, the removal was quite close for both catalytic systems. Nonetheless, the percent photolysis under UV-A was 1.2%, whereas under visible light was negligible. Based on the above-mentioned results, the as-synthetized FeC_2O_4-based compound exhibited photocatalytic activity under UV-A and visible irradiation that led to removal of the alkaloid caffeine.

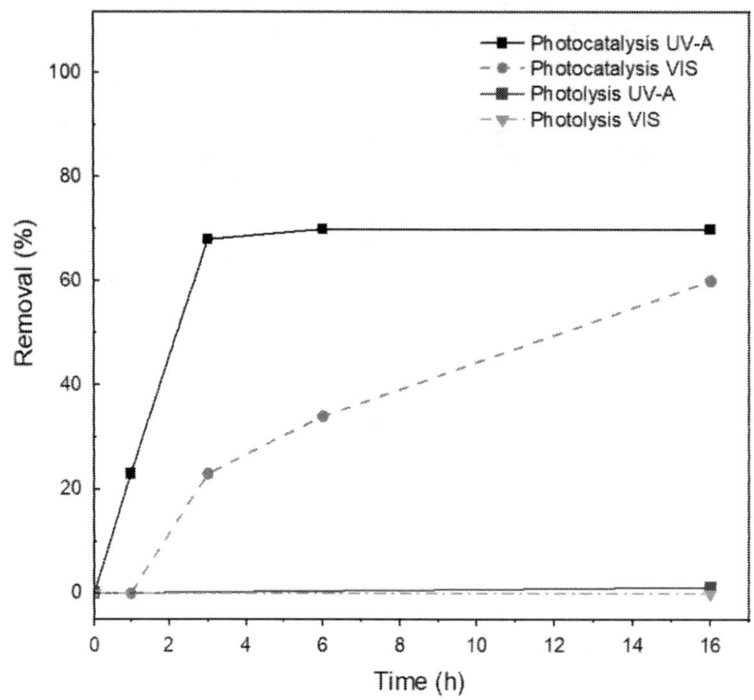

Figure 11. Caffeine percent removal by photolysis and photocatalysis via the FeC_2O_4-based synthetized catalyst

4. Conclusions

The synthesis of Fe/Ti-based compounds from ferro-titaniferous mineral sands using a non-hazardous solvent system (1 M oxalic acid) at subcritical water conditions (135 °C and 4 h) was demonstrated as a feasible alternative to conventional mineral acid leaching. The as-synthetized compound was identified as ferrous oxalate hydrate by XRD. The synthetized ferrous oxalate hydrate exhibited photocatalytic activity under both UV-A and visible light irradiation as methylene blue and caffeine were indeed photodegraded. Therefore, the synthetized ferrous oxalate can be potentially used as a photocatalyst to effectively utilize the solar energy in advanced oxidation processes for wastewater treatment.

Acknowledgements

This research was supported by Escuela Politécnica Nacional (EPN) through the project grant PIJ-18-01. The authors would also like to thank the Electroceramics Laboratory of the Department of Physics (EPN), and the New Materials Laboratory of the Department of Materials (EPN) for material characterization, as well as the FSPI Project - Doctoral Schools, of the French Embassy in Ecuador, financed by the Ministry of Europe and Foreign Affairs.

References

[1] Antonopoulou M, Kosma C, Albanis T and Konstantinou I 2021 An overview of homogeneous and heterogeneous photocatalysis applications for the removal of pharmaceutical compounds from real or synthetic hospital wastewaters under lab or pilot scale *Sci. Total Environ.* **765** 144163

[2] Khan S, Naushad M, Al-Gheethi A and Iqbal J 2021 Engineered nanoparticles for removal of pollutants from wastewater: Current status and future prospects of nanotechnology for remediation strategies *J. Environ. Chem. Eng.* **9** 106160

[3] Mashile G P, Selahle S K, Mpupa A, Nqombolo A and Nomngongo P N 2022 Chapter 8 - Remediation of emerging pollutants through various wastewater treatment processes *Emerging Freshwater Pollutants* ed T Dalu and N T Tavengwa (Elsevier) pp 137–50

[4] Negrete-Bolagay D, Zamora-Ledezma C, Chuya-Sumba C, De Sousa F B, Whitehead D, Alexis F and Guerrero V H 2021 Persistent organic pollutants: The trade-off between potential risks and sustainable remediation methods *J. Environ. Manage.* **300** 113737

[5] Kumar L, Chugh M, Kumar S, Kumar K, Sharma J and Bharadvaja N 2022 Remediation of petrorefinery wastewater contaminants: A review on physicochemical and bioremediation strategies *Process Saf. Environ. Prot.* **159** 362–75

[6] Coha M, Farinelli G, Tiraferri A, Minella M and Vione D 2021 Advanced oxidation processes in the removal of organic substances from produced water: Potential, configurations, and research needs *Chem. Eng. J.* **414** 128668

[7] Xie Z H, Zhou H Y, He C S, Pan Z C, Yao G and Lai B 2021 Synthesis, application and catalytic performance of layered double hydroxide based catalysts in advanced oxidation processes for wastewater decontamination: A review *Chem. Eng. J.* **414** 128713

[8] Ameta R, Chohadia A K, Jain A and Punjabi P B 2018 *Fenton and Photo-Fenton Processes*

[9] Miklos D B, Remy C, Jekel M, Linden K G, Drewes J E and Hübner U 2018 Evaluation of advanced oxidation processes for water and wastewater treatment – A critical review *Water Res.* **139** 118–31

[10] Es-sahbany H, El Yacoubi A, El Hachimi M L, Boulouiz A, Chafik El Idrissi B and El Youbi M S 2022 Low-cost and eco-friendly Moroccan natural clay to remove many bivalent heavy metal ions: Cu2+, Co2+, Pb2+, and Ni2+ *Mater. Today Proc.* 1–7

[11] Ramirez-Muñoz A, Pérez S, Muñoz-Saldaña J, Flórez E and Acelas N 2021 Eco-friendly materials obtained through a simple thermal transformation of water hyacinth (Eichhornia Crassipes) for the removal and immobilization of Cd2+ and Cu2+ from aqueous solutions *Environ. Nanotechnology, Monit. Manag.* **16**

[12] Lessa E F, Nunes M L and Fajardo A R 2018 Chitosan/waste coffee-grounds composite: An

efficient and eco-friendly adsorbent for removal of pharmaceutical contaminants from water *Carbohydr. Polym.* **189** 257–66

[13] Kumar B, Smita K, Galeas S, Guerrero V H, Debut A and Cumbal L 2021 One-Pot Biosynthesis of Maghemite (γ-Fe2O3) Nanoparticles in Aqueous Extract of Ficus carica Fruit and Their Application for Antioxidant and 4-Nitrophenol Reduction *Waste and Biomass Valorization* **12** 3575–87

[14] Rahim Pouran S, Abdul Raman A A and Wan Daud W M A 2014 Review on the application of modified iron oxides as heterogeneous catalysts in Fenton reactions *J. Clean. Prod.* **64** 24–35

[15] Kim S R, Kim S and Kim E J 2020 Photoreaction characteristics of ferrous oxalate recovered from wastewater *Chemosphere* **249** 126201

[16] Hu L, Wang P, Xiong S, Chen S, Yin X, Wang L and Wang H 2019 The attractive efficiency contributed by the in-situ reactivation of ferrous oxalate in heterogeneous Fenton process *Appl. Surf. Sci.* **467–468** 185–92

[17] Wang G, Zhou A and Xu Q 2019 α-Ferrous oxalate with different micro scale: Synthesis and catalytic degradation effect to rhodamine B *Solid State Sci.* **91** 54–60

[18] Vedrenne M, Vasquez-Medrano R, Prato-Garcia D, Frontana-Uribe B A, Hernandez-Esparza M and de Andrés J M 2012 A ferrous oxalate mediated photo-Fenton system: Toward an increased biodegradability of indigo dyed wastewaters *J. Hazard. Mater.* **243** 292–301

[19] Kwan C Y, Chu W and Lam W S 2007 The role of oxalate in the kinetics of 2,4-D oxidation over ferrous ion-supported catalysts *J. Mol. Catal. A Chem.* **274** 50–7

[20] Liu L, Chen Z, Zhang J, Shan D, Wu Y, Bai L and Wang B 2021 Treatment of industrial dye wastewater and pharmaceutical residue wastewater by advanced oxidation processes and its combination with nanocatalysts: A review *J. Water Process Eng.* **42** 102122

[21] Lin K Y A and Chen B J 2017 Magnetic carbon-supported cobalt derived from a Prussian blue analogue as a heterogeneous catalyst to activate peroxymonosulfate for efficient degradation of caffeine in water *J. Colloid Interface Sci.* **486** 255–64

[22] Lagos K J, Marinkovic B A, Dosen A, Guamán M V., Guerrero V H, Pardo E and Pontón P I 2020 Data on phase and chemical compositions of black sands from "El Ostional" beach situated in Mompiche, Ecuador *Data Br.* **32** 106214

[23] Valdivieso-Ramírez C S, Pontón P I, Dosen A, Marinkovic B A and Guerrero V H 2022 One-Step Synthesis of Iron and Titanium-Based Compounds Using Black Mineral Sands and Oxalic Acid under Subcritical Water Conditions *Minerals* **12** 306

[24] Liu S, Wang X and Zhang J 2020 Fenton-like reaction and photocatalysis using ferrous oxalate and G-C3 N4 enhancing reactive oxygen species for dye wastewater degradation under visible-light irradiation *Desalin. Water Treat.* **193** 359–68

[25] Frost R L 2004 Raman spectroscopy of natural oxalates *Anal. Chim. Acta* **517** 207–14

[26] Zhang T, Oyama T, Aoshima A, Hidaka H, Zhao J and Serpone N 2001 Photooxidative N-demethylation of methylene blue in aqueous TiO2 dispersions under UV irradiation *J. Photochem. Photobiol. A Chem.* **140** 163–72

[27] Kwan C, Technology W C-W S and and 2004 undefined 2004 Photooxidation of 2, 4-dichlorophenoxyacetic acid by ferrous oxalate-mediated system *iwaponline.com*

[28] Kolayli S, Ocak M, Küçük M and Abbasoğlu R 2004 Does caffeine bind to metal ions? *Food Chem.* **84** 383–8

[29] Nafisi S, Sadjadi A S, Zadeh S S and Damerchelli M 2003 Interaction of metal ions with caffeine and theophylline: Stability and structural features *J. Biomol. Struct. Dyn.* **21** 289–95

[30] Jabbar M A, Kabir M H and Chowdhury R A 2013 Influence Of Caffeine On The Redox Characteristics Of Iron In Aqueous Solution *J. Bangladesh Chem. Soc.* **25** 166–79

[31] Calatayud J M and Zamora L L 2013 Spectrophotometry | Pharmaceutical Applications *Encycl. Anal. Sci.* 249–62

[32] Dalga B and Belay A 2017 Study of the self-association of caffeine and chlorogenic acid and their hetero-association with methylene blue using spectrophotometric method *rjb.ro* **27** 35–53

[33] Dalmázio I, Santos L S, Lopes R P, Eberlin M N and Augusti R 2005 Advanced oxidation of caffeine in water: On-line and real-time monitoring by electrospray ionization mass spectrometry *Environ. Sci. Technol.* **39** 5982–8

[34] Rao Y, Long H and Hao J 2021 The oxidative degradation of Caffeine in UV/Fe(II)/persulfate system—Reaction kinetics and decay pathways *Water Environ. Res.* **93** 559–69

[35] Norazmi N, Rasad Z R A, Mohamad M and Manap H 2017 Uric acid detection using uv-vis spectrometer *IOP Conf. Ser. Mater. Sci. Eng.* **257**

[36] Nuevo M, Milam S N, Sandford S A, Elsila J E and Dworkin J P 2009 Formation of uracil from the ultraviolet photo-irradiation of pyrimidine in pure H2O ices *Astrobiology* **9** 683–95

Polarization and biomineralization of hydroxyapatite-barium titanate composites

R Uribe[1*], I Rojas[1], M C Riofrio[1], L Lascano[2] and G González[3-4*]

[1]Escuela Politécnica Nacional, Dpto. de Ingeniería Química. PO Box 17-01-2759, Quito-Ecuador.
[2]Escuela Politécnica Nacional, Dpto. de Física. PO Box 17-01-2759, Quito-Ecuador.
[3]School of Physics and Nanotechnology, Yachay Tech University, Urcuquí 100650, Ecuador.
[4]Instituto Venezolano de Investigaciones Científicas, Centro de Ingeniería Materiales y Nanotecnología, Caracas 1020-A, Venezuela.
* Correspondence: rafael.uribe@epn.edu.ec and ggonzalez@yachaytech.edu.ec

Abstract. In the search of new materials for bone regeneration, the materials with piezoelectric properties look very promising. It has been reported that piezoelectric materials induce bone growth and enhance implant integration. Additionally, it has been found that bioactivity increases in negatively charge surfaces such as polarized $BaTiO_3$ (BT) and polarized hydroxyapatite (HAp). Additionally, it is known that $BaSO_4$ (BS) increases biocompatibility. Therefore, in this work, composites materials of 80BT/20BS (BTS) in different proportions with HAp (HAp/BTS: 10/90, 30/70, 50/50, 70/30 y 90/10) were prepared by a mixture of nanometric powders and then they were sintered at 1000 °C for a period of 5 h. The materials were polarized at 130, 300 and 400 °C applying a DC electric field of 1 kV/mm, during 1 h. The electric field was maintained until the material was cooled down to room temperature. The electric and piezoelectric response were measured immediately after cooling, after 1 h and after 24 h. The dielectric measurements of materials were performed at different frequencies (0.1 to 100 kHz). The polarized and unpolarized materials were immersed in simulated body fluid (1.5 SBF) for 7 and 19 days. The deposition and growth of hydroxyapatite using the biomimetic method was followed by FTIR and SEM. The polarization effect on the crystalline growth of hydroxyapatite formed from the SBF solution has been demonstrated. The process of biomineralization of HAp on HAp/BTS composites increased considerably with the addition of barium titanate, this effect greatly improved in polarized materials. The typical coral-like morphology characteristic of HAp formation from SBF deposition was observed after 7 days of SBF immersion for polarized composites. Excellent dielectric properties were determined by adding 30% Of $BaTiO_3$, obtaining for these composites dielectric constant values of the order of 20 to 10 kHz, values similar to that of human bones. Therefore, these materials look very promising for bone regeneration.

Keywords: Polarization, biomaterials, barium titanate, hydroxyapatite, biomineralization.

1. Introduction

The need to develop new materials with outstanding properties for bones implants is a very important field of research. For many years, hydroxyapatite (HAp) has been employed as a biomaterial, mainly due to its similarity to the mineral phase of bone tissue. However, HAp is a bioceramic material with low tensile strength, and its clinical applications have been limited to uses in case where no high mechanical stress is applied. There is an interest on obtaining HAp with mechanical properties as closely as possible to those of natural bones, which would allow an adequate bone regeneration at the implant site [1-10]. For several years, HAp has been used in bone regeneration implants mainly due to its excellent biological properties (biocompatibility, bioaffinity, bioactivity, osteoconduction, osseointegration and osteoinduction) [11]. Likewise, HAp surface withstands adhesion, growth and differentiation of osteoblasts cells, which form a new bone by progressive replacement of the adjacent bone. In this way, the formation of an apatite layer allows an implant to bind directly to the host tissue [12,13]. In the last years, there has also been interest in the study of electrically charged materials, due to the electrical activity of bones. Thus, how to increase the applications and properties of HAp, like those related to the piezoelectric properties, has been investigated [9-10]. Both hydroxyapatite and collagen have weak piezoelectric properties, which are of great importance for bone regeneration [14-16]. The bone tissue is made up of a 60-70 % HAp, 5-8 % water, 30-40 % of organic matter (mainly collagen) and 5-8 % of other proteins. Its mechanical properties are Young's modulus (modulus of elasticity) of 17-20 GPa, tensile strength of 120-150 GPa and compressive strength of 100-160 GPa. Therefore, developing a material that combines together these properties for being used as a replacement, constitute a challenge. Recent studies show that dry bone is piezoelectric, whereas others state that fully hydrated bone does not have good piezoelectric properties for all physiological frequencies. The hypothesis that these properties have a positive influence on bone regeneration is of interest. The bone piezoelectric properties are the result of the collagen fibers movement under a mechanical load. The electric dipole in bones is formed because of asymmetric collagen molecules, which are parallel oriented to the bone axis [10,17-21]. An important property to highlight is that the conductivity of hydroxyapatite in air goes from 10^{-9} to 10^{-7} S/cm for a temperature range between 500 °C and 800 °C. This is important because, in the presence of an external electric field, the HAp protons are displaced without altering the crystalline symmetry by raising the temperature [22]. Furthermore, it has been shown that polarized HAp has a significant effect on the crystallization of dissolved inorganic ions *in vitro* [10]. The effect of a charge surface development on HAp discs concerning osteoblast activity *in vitro* was studied by Kumar et al. [23]. They have sintered discs with and without steam, from 4 h to 7 days. They concluded that all tested discs supported cell viability and binding. In this way, cell binding, proliferation and metabolic activity showed a significant increase at higher sintering time with polarized samples. On the other hand, barium titanate ($BaTiO_3$) is a highly studied ferroelectric material due to its simple crystalline structure, chemical and mechanical stability, good ferroelectric properties and a high dielectric constant [24-25]. However, there are few studies as a biomaterial despite its excellent piezoelectric properties. $BaTiO_3$ has a high dielectric constant (CD), approximately 6000 for a fine grain size (< 1 μm), and between 1500 and 2000 for a coarse grain size (> 1 μm). Its Curie temperature is between 120 °C and 130 °C, which is an important parameter because bellow this temperature the material shows ferroelectric properties [26-27]. There has been an interest to develop electrically active ceramics to be used in hard tissues, providing a biological response improvement, since it has been established that such material is biocompatible and improves bone growth [10,18]. Park et al. [28] studied $BaTiO_3$ as a direct substitute in hard tissues, electrically stimulated (polarized) and inactive (unpolarized). They made implants and tested them in the femur of several dogs for 95 days. Their results indicated a high degree of biocompatibility, with neither inflammatory response nor a rare response at the tissue-body interphase. These authors did not find any difference between polarized and no polarized implants. Recent studies have demonstrated that piezoelectric properties increase bioactivity when combining barium titanate with hydroxyapatite, for $BaTiO_3$-HAp compounds [29-31]. Also, some studies have proposed biomineralization, to simulate hydroxyapatite formation in the human body, using a simulated body fluid (SBF), predicting the *in vivo* bone bioactivity of a material [32]. As mentioned, natural bone has a piezoelectric behavior, that inspires research to add a piezoelectric component to the bone implant to obtain favorable biological responses. Several piezoelectric ceramics have been investigated, among them $BaTiO_3$ and lead zirconium titanate (PZT). Nevertheless, the latter, despite its good piezoelectric properties, has a high degree of toxicity when it is employed as a bone substitute. For this reason, $BaTiO_3$ is the one of greatest interest [33]. HAp is a nontoxic biomaterial and, lately, improvements for bone integration have been investigated, being polarization a good option. Some studies show how both electrical polarization and temperature influence the growth bone-like crystals on HAp ceramics Yamashita et al. [34]. Piezoelectric $BaTiO_3$ combined with HAp looks as a material that, in addition to having a chemical similarity with the bone mineral content, would also respond to mechanical stress. The piezoelectric effect of BT has also been observed to promote growth of osteoblasts on HAp-BT compounds [35]. Likewise, optical and dielectric properties of thick films of the HAp-$BaTiO_3$ compound have been studied, and it has been shown that an increase of HAp concentration leads to a decrease in the film's density and dielectric constant [26]. On the other hand, Bowen et al. [30] measured the dielectric and piezoelectric properties of compounds, changing $BaTiO_3$ compositions from 0 to 100 %. Their results show that the addition of $BaTiO_3$, for compositions over 80 % increased the piezoelectric properties of the material. The reports of Baxter et al. [10] reveled that HAp-$BaTiO_3$ compounds with less than 70 % of $BaTiO_3$ did not present measurable piezoelectric properties. Later, Baxter et al. [18] found that piezoelectric coefficient of ceramics containing 90% of $BaTiO_3$ obtained a better response. This study becomes a baseline because they demonstrated with *in-vitro* studies that HAp-$BaTiO_3$ compounds are biocompatible. Panda et al. [31] reveled an increase of relative permittivity and a decrease in dielectric dissipation with increase in the amount of barium titanate in the HAp-$BaTiO_3$ compound. Consequently, a higher amount of barium titanate may improve hydroxyapatite polarization due to the permittivity rise and reduced dissipation values. Compounds of HAp-$BaTiO_3$ with porosities of 40, 50 y 60 %, with and without polarization treatment, were prepared by Liu et al. [36]. Their results showed that the more porous and polarized compounds the better the performance regarding the adherence and osteoblast proliferation. Prakasam et al. [37] analyzed HAp-$BaTiO_3$ compound's behavior and the influence of grain size (300 and 500 nm) and stated that increase of sintering temperature and BT could cause the HAp dissociation, which leads to have different phases such as: β-TCP / α-TCP / $CaTiO_3$/ $Ba_{10}(CPO_4)_6(OH)_2$ and loss of ions in HAp. On the other hand, $BaSO_4$ particles have been studied in bone cements resulting in increased biocompatibility, improved mechanical properties and significant increase of osteoblast proliferation [38].

In accordance with the above, it is necessary to develop suitable materials for bone implants. In this regard, the combination of hydroxyapatite, barium titanate and barium sulfate could give excellent results, being a research field with scarce studies. Therefore, in this paper we develop a new composite of HAp/($80BaTiO_3$-$20BaSO_4$), HAp-BTS, based on the control composition, sintering temperature, polarization parameters and their effect on the biomineralization process. As far as the authors knowledge up to now there are not reports on this new compound and the study of its biomineralization behavior after polarization.

2. Materials and methods

HAp-BTS compounds (Serie: 100 % HAp - 10/90 - 30/70 - 50/50 -70/30 - 90/10 - 100 % BTS) were prepared from the mixture of commercial carbonated hydroxyapatite powder (99.99 % purity, from Sudeep Pharma, India), barium titanate and barium sulfate (99.99 % purity, from Sigma Aldrich). BTS proportion corresponds to the mixture of 80% $BaTiO_3$ - 20% $BaSO_4$. Compounds were prepared in triplicate by high-speed mechanical stirring (Ultraturrax-T25 method) in distilled water for 5 minutes, dried at 80 °C for 6 hours, pressed on 12 mm diameter discs at 10000 psi and sintered at 1000 °C for 5 hours.

Polarization

For polarization measurement, the sintered material is placed between two platinum electrodes, which are used to apply the electric field [28]. The discs coated with silver electrodes were heated to temperatures of 120 °C,130 °C, 300 °C and 400 °C, to facilitate the movement of the domains, applying an electric field of 1 kV/mm for 1 hour. Finally, the material was carried out to room temperature while the field has still been applied.

Dielectric constant

The composite conductance and capacitance values were measured, for an average of 16 measurements per sample, using a LCR Meter 4263B at frequencies between 100 Hz and 100 kHz. The determination of dielectric constant was obtained using the equation: [39]

$$\varepsilon' = \frac{Cd}{\varepsilon_0 A} \quad (1)$$

where:
ε': dielectric constant
C: capacitance
d: sample thickness
ε_0: vacuum dielectric constant (8.85×10^{-12} F.m^{-1})
A: sample surface

Composites biomineralization
In order to test the bioactivity of the compounds immersion in a simulated body fluid (SBF) was carried out, for predicting *in vitro* bone bioactivity. SBF solution has ions concentrations almost equal to those of human blood plasma, a 1.5 SBF was prepared [32,40]. Before SBF immersion, the discs were washed with alcohol, in an ultrasound bath for 5 minutes, then dried with compressed air and weighed. The substrates were immersed for 7 and 19 days in a SBF volume calculated according to equation (2), at 37 °C and a solution pH between 7.2 and 7.4:

$$V_s = {}^{S_a}/_{10} \quad (2)$$

where:
V_s: simulated body fluid volume (mL)
S_a: substrate surface area (mm^2)

Composites characterization
Characterization of the HAp and BTS powders, as well as the composites and the HAp films formed on the immersed composites in the SBF for 7 and 19 days, was carried out by scanning electron microscopy (SEM), elemental analysis (EDS) and Fourier transforms infrared spectroscopy (FTIR), the latter with a Perkin Elmer Spectrum 100 spectrophotometer in a range of 4000-500 cm^{-1}.

3. Results and Discussion

Composites HAp/BTS structural characterization
Figure 1 shows the FTIR spectra of HAp and BTS powders and sintered material at 1000 °C - 5 h. The characteristic bands of phosphates PO_4^{-3}- ν_3 at 1090 cm^{-1} and 1040 cm^{-1}, and phosphate (PO_4^{-3}- ν_1) at 962 cm^{-1}, can be observed in figure 1a, these are considered HAp fingerprints. Both HAp curves present a PO_4^{-3}- ν_2 phosphate groups at 450 cm^{-1} and PO_4^{-3}- ν_4 at 600 cm^{-1} and 550 cm^{-1}. Furthermore, non-sintered HAp spectra reveal characteristic bands of OH^- structural and (CO_3^{-2}- ν_2) carbonate groups, that disappear after sintering process. A broad band characteristic of $BaTiO_3$ from 800 to 450 cm^{-1} is shown in figure 1b. The latter is centered around 578 cm^{-1} corresponding to (Ti-O) stretching. Likewise, bands corresponding to the vibrations of SO_4^{2-} at 1116 cm^{-1}, 1075 cm^{-1} and 987 cm^{-1} are observed [41-42].

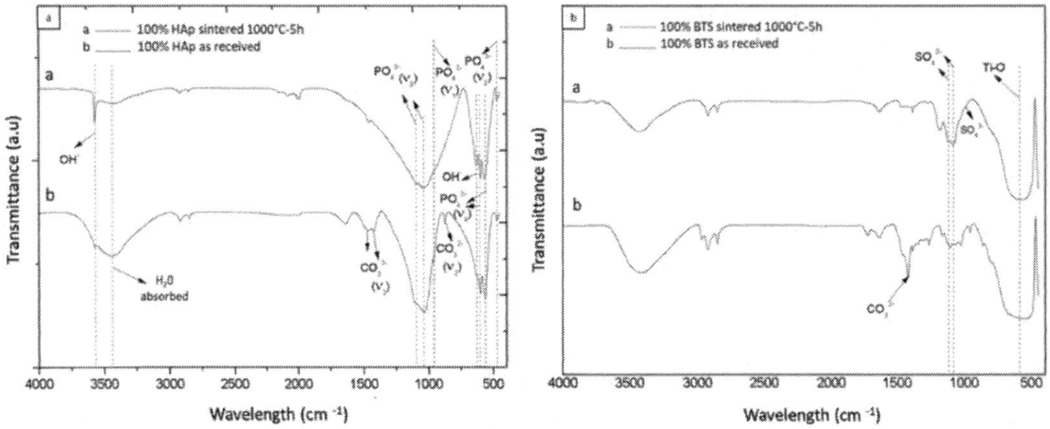

Figure 1. FTIR spectra of HAp and BTS as received and composite sintered at1000°C - 5h. (a) HAp and (b) BTS.

HAp-BTS composites sintered at 1000 °C - 5h FTIR spectra are presented in figure 2. It can be seen that the HAp and BTS characteristic bands are present in all samples but with different intensities. In samples with more HAp proportion, the characteristics band of PO_4^{-3}- ν_4 phosphate group at 1040 cm^{-1} show increase intensity [37].

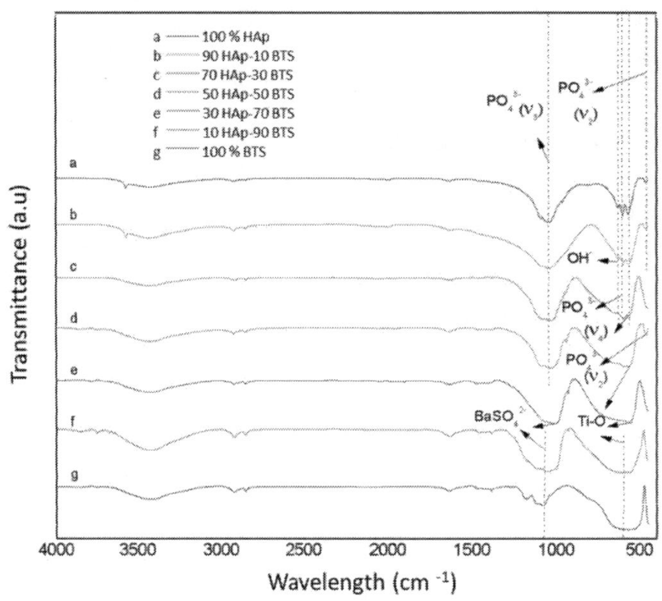

Figure 2. FTIR spectra of composites HAp/BTS sintered at 1000°C - 5h.

The SEM images and the distribution of grain size for sintered HAp/BTS composites at 1000 °C - 5h are shown in figure 3. The images 3a, 3b and 3c show the influence of a higher proportion of BTS mixture in

 2238 (2022) 012007 doi:10.1088/1742-6596/2238/1/012007

the microstructure development, providing evidence of the formation of some phase mixtures with low melting point. These results corroborate the findings of Vouilloz et al. [43], they studied $BaTiO_3$-HAp compounds reactivity and stated that secondary phase formation and HAp transformation into tricalcium phosphates depend on the reactivity of the starting materials, they also found secondary phases of calcium ($Ca_2P_2O_7$; $CaTiO_3$) and traces of $Ba_3(PO_4)_2$, concluding that the formation of these phases facilitates the release of ions into the incubation solution thus biocompatibility, depending on sample composition.
From SEM images and using *J image* processing software, the histograms shown in figures 3d, 3e and 3f were obtained. These images correspond to grain diameter distribution for each sample, the main grain size for composites is 1.2 μm presenting a slight size increase for 30HAp-70BTS composite.

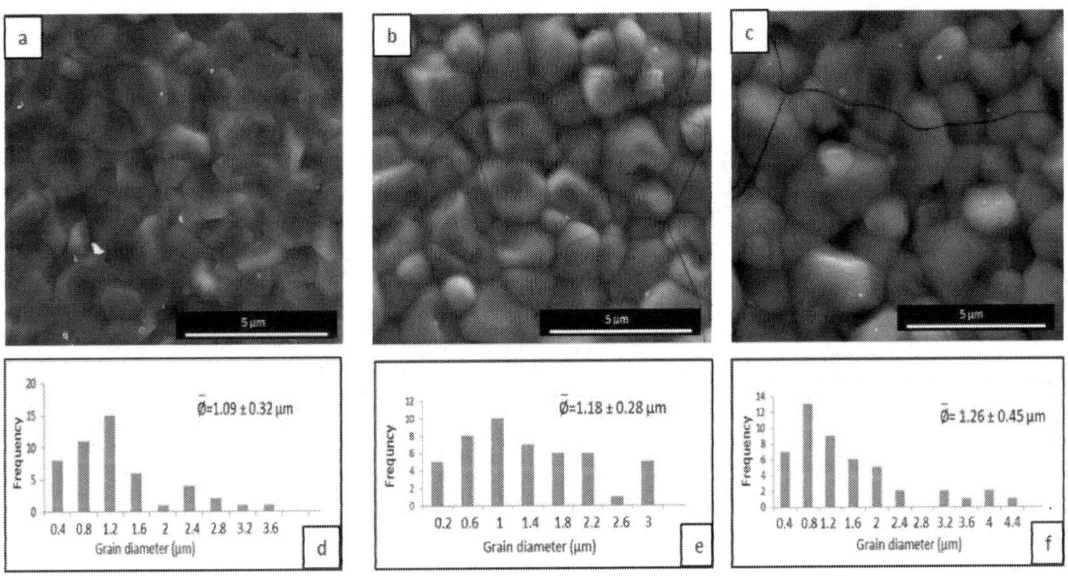

Figure 3. SEM images of composites HAp-BTS sintered a 1000°C - 5h: a) 90HAp-10BTS, b) 70HAp-30BTS and c) 30HAp-70BTS.

Dielectric property evaluation
Table 1 provides the dielectric constant values for HAp-BTS sintered composites measured in the range of 100 Hz and 100 kHz. The dielectric constant as a function of frequency for the different compositions is plotted in figure 4. First of all, it is observed that the dielectric constant decrease as the frequency increases. Second, for each frequency, the dielectric constant decrease significantly as HAp content increases, for instance, for 100 Hz the dielectric constant of 10HAp-90BTS composite decreases approximately 40 % with a 10% of HAp in comparison with dielectric constant of 100 BTS. The effective dielectric constant can be reduced due to: grain structure, lack of connectivity between phases, formation of sintered reaction products and porosity. Dubey et al. [44] reported excellent dielectric properties adding 40% of $BaTiO_3$, and their mixtures had similar characteristics to bone, with dielectric constant of 20.8 for 10 kHz, these values are similar to ours for 70HAp-30BTS composite. The piezoelectric coefficient for human bone is 0.7 pC/N [44], which is consistent in magnitude with the value of (≈2.6 pC/N) found for the compound 70HAp–30BTS, demonstrating that BTS mixing improves piezoelectric properties significantly. A high piezoelectric coefficient (≈90 pC/N) was obtained for 100 BTS compound.

Table 1. Dielectric constant of sintered composites HAp-BTS.

Composites	Dielectric constant					Piezoelectric Coefficient (pC/N)
	100 Hertz	120 Hertz	1 kHertz	10 kHertz	100 kHertz	
100 HAp	24.1	22,4	16,1	13.0	10.8	-
90 HAp - 10 BTS	97.3	86.3	29.9	16.0	12.3	-
70 HAp - 30 BTS	117.6	108.0	44.8	21.7	14.5	2.6
50 HAp - 50 BTS	137.5	128.7	59.2	29.3	21.6	9
30 HAp - 70 BTS	288.0	274.9	141.5	58.5	36.6	23
10 HAp - 90 BTS	597.4	574.0	291.1	179.6	153.3	52
100 BTS	936.1	924.5	835.9	798.1	771.1	90

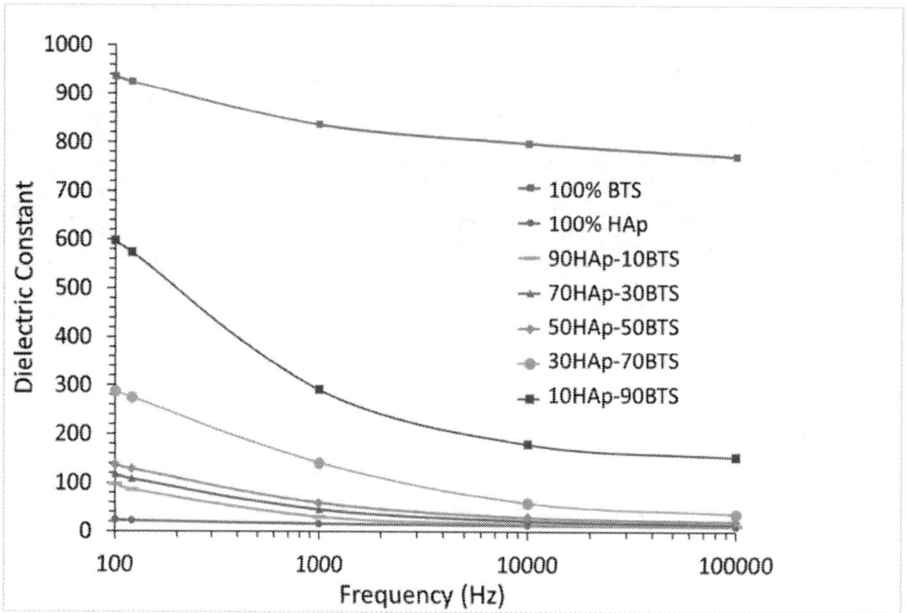

Figure 4. Dielectric constant composites HAp-BTS sintered at 1000°C - 5h.

Biomineralization

Figure 5 shows both SEM images of HAp deposition on unpolarized and polarized 100 HAp material, after an immersion for 7 days into SBF solution, as well as their FTIR spectra. Uniform microstructure with rounded grains, characteristic of pure HAp is observed in figure 5a. Figure 5b, corresponds to the unpolarized biomineralized composite, a deposition in coral shape is observed typical of apatite deposition after SBF immersion. Whereas for polarized samples (figure 5c) the film obtained after SBF immersion is more uniform and in the form of a rounded particles. Likewise, from the EDS analysis it has been obtained a Ca/P ratio of 1.47 and 1.61 for unpolarized and polarized composites respectively, these indicates the formation of apatite or non-stoichiometric HAp. These results reinforce the feasibility of use pure HAp for biomedical applications. It has been reported by Yamashita et al. [34] that in samples immersed into SBF, cations are absorbed faster on polarized surfaces than on unpolarized ones. FTIR spectra (figure 5d) shows presence of apatite deposits, due to the characteristic band of PO_4^{-3}- ν_3 at 1020 cm^{-1}.

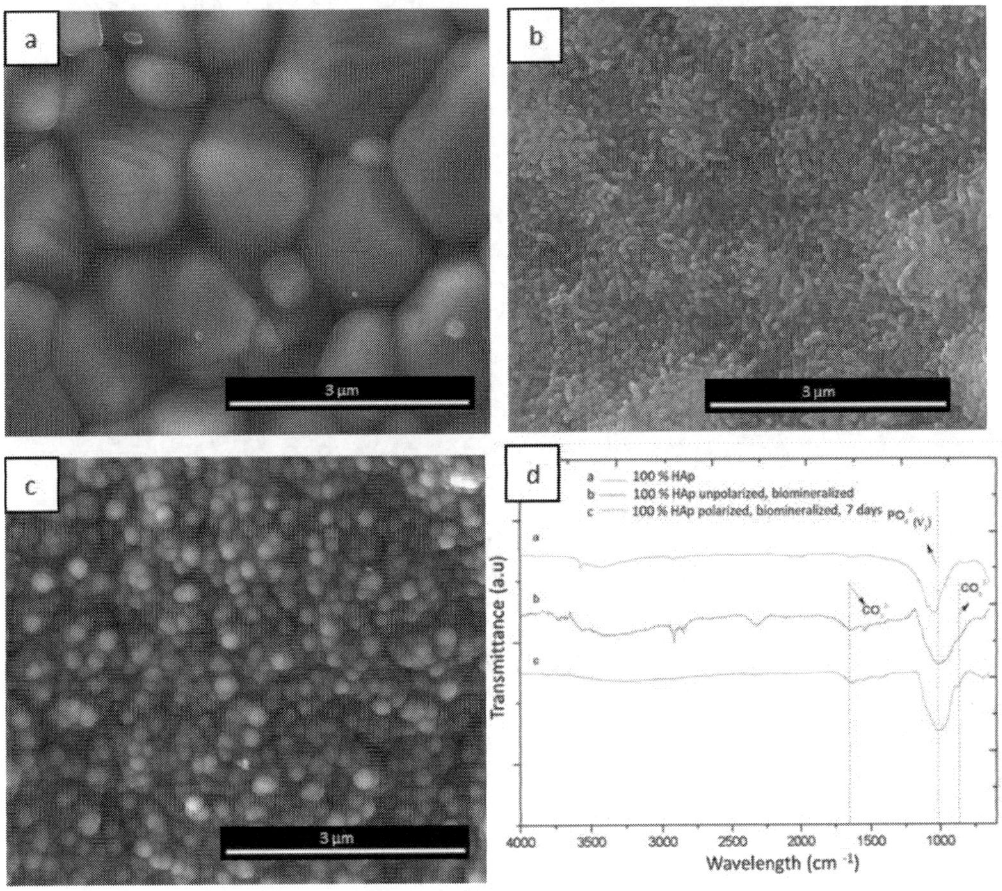

Figure 5. SEM images of composite sintered 100 % HAp after immersion in SBF: a) microstructure of sintered material, b) unpolarized after immersion in SBF for 7 days, c) polarized after immersion in SBF 7 days and d) FTIR of different samples.

Figure 6 present SEM images of HAp deposition on 50HAp-50BTS composite unpolarized and polarized, immersed for 7 and 19 days in SBF solution. In figure 6a show a uniform microstructure with rounded grains characteristic of sintered HAp composites. The typical coral shape morphology characteristic of apatite formation from SBF deposition is observed in figure 6b after 7 days of SBF immersion for the unpolarized material and in figure 6c for the polarized material. The growth and deposition of HAp crystals in polarized composites is higher and better defined after 19 days of immersion (figure 6d). Furthermore, the EDS analysis reveals the formation of stoichiometric HAp with Ca/P ratio of 1.67 for the 19 days biomineralized samples (figure 6e). It is demonstrated that the formation of the HAp film after SBF immersion is enhanced with a polarized BTS mixture, indicating the material improvement in bioactive character after polarization.

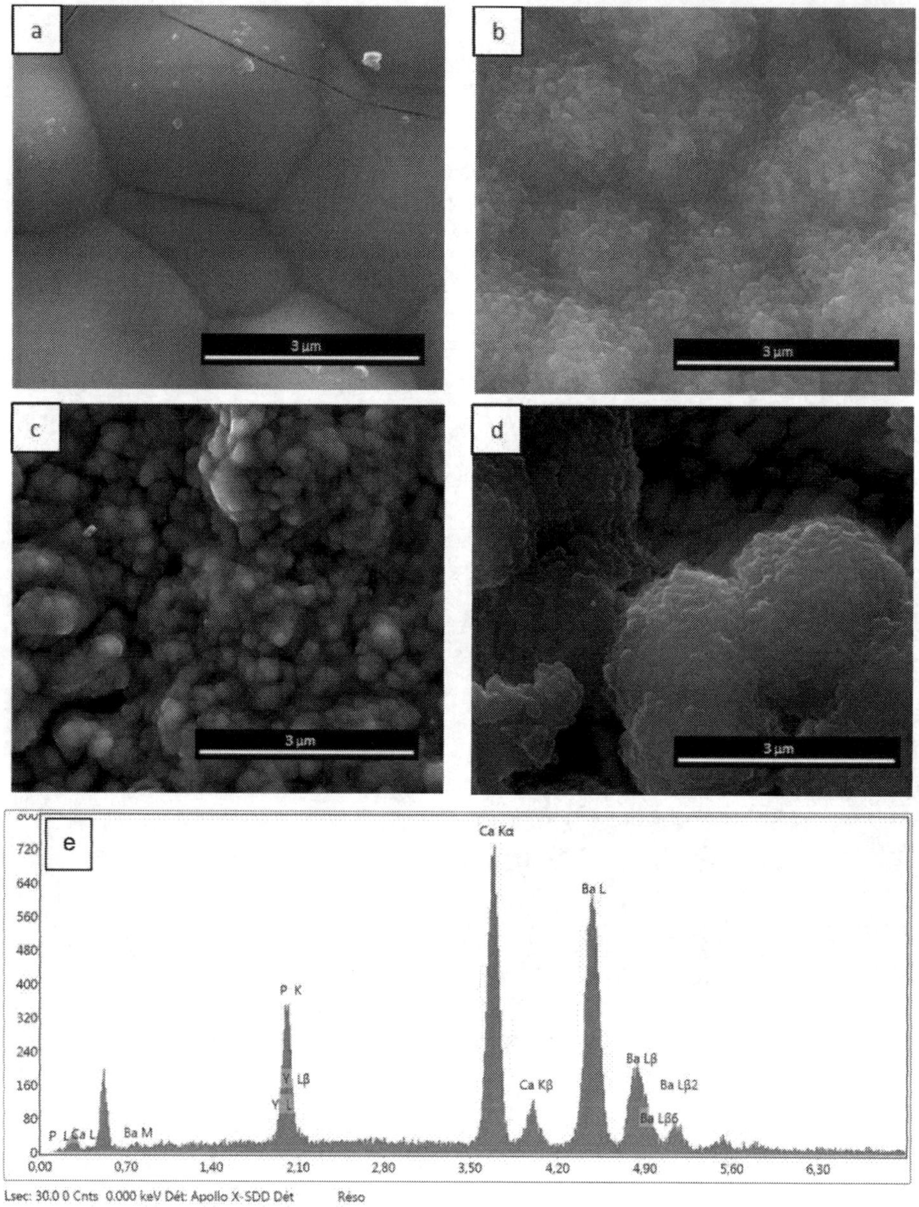

Figure 6. SEM images of composite 50HAp-50BTS sintered at 1000°C - 5h: a) microstructure of sintered material, b) unpolarized after immersion in SBF for 19 days, c) polarized after immersion in SBF 7 days, d) polarized after immersion in SBF for 19 days and e) EDS polarized after immersion in SBF for 19 days

4. Conclusions

Different mixtures of HAp-BTS have been successfully prepared with an uniform grain distribution and interesting dielectric properties, similar to the dielectric properties of bone. The addition of BTS to HAp composites increase the piezoelectric properties of hydroxyapatite and after polarization of the composites their bioactivity was significantly increased. This result suggests that this new compounds HAp-BTS could potentially be used for bone regeneration with increase biocompatibility and bioactivity.

Funding: This research was funded by the Project EPN-PIMI 17-05: "Desarrollo de sistemas de hidroxiapatita dopados con óxidos metálicos biocompatibles para aplicaciones biomédicas". Escuela Politécnica Nacional, Quito-Ecuador.

Acknowledgments: The authors would like to thank Vicerrectorado de Investigación, Innovación y Vinculación, Escuela Politécnica Nacional and Universidad Yachay Tech del Ecuador.

References

[1] Martínez A, Esparza H y Ortiz J **2008** Caracterización estructural y morfológica de hidroxiapatita nanoestructurada: estudio comparativo de diferentes métodos de síntesis. *Superficies y vacío*, **21(4)**, 18-21.

[2] Hernández T, Palma C y Pina B **1999** Hydroxyapatite and its applications; Hidroxiapatita y sus aplicaciones. *Revista mexicana de física*, ***45***.

[3] Sequeda L, Díaz J, Gutiérrez S J, Perdomo S y Gómez O **2012** Obtención de hidroxiapatita sintética por tres métodos diferentes y su caracterización para ser utilizada como sustituto óseo. *Revista Colombiana de Ciencias Químico-Farmacéuticas*, ***41*(1)**, 50-66.

[4] García M y Reyes J **2006** La hidroxiapatita, su importancia en los tejidos mineralizados y su aplicación biomédica. *TIP Revista especializada en ciencias químico-biológicas*, ***9*(2)**, 90-95.

[5] Montoliu F **1993** Biomateriales I. Clasificaciones de los materiales. *Revista de la Reial Acadèmia de Medicina de Catalunya*, 139-151.

[6] Park J y Bronzino J **2002** *Biomaterials: principles and applications*. CRC Press.

[7] Velasco M y Garzón D **2010** Implantes Scaffolds para regeneración ósea.: Materiales, técnicas y modelado mediante sistemas de reacción-difusión. *Revista Cubana de Investigaciones Biomédicas*, ***29*(1)**, 0-0.

[8] Soria J, Barcia J, Andrades A, Romero J, Monleón M y García J **2009** Uso de biomateriales en medicina regenerativa, aspectos básicos y aplicaciones en el Sistema Nervioso. *Trauma (Majadahonda)*, 15-22.

[9] Martinelli A, Novoa M, Oldani C y Corominas A **2011** Síntesis y caracterización de hidroxiapatita para implantes biomédicos. In *XVIII Congreso Argentino de Bioingeniería SABI*.

[10] Baxter R, Turner G, Bowen R, Gittings P y Chaudhuri J **2009**. An in vitro study of electrically active hydroxyapatite-barium titanate ceramics using Saos-2 cells. *Journal of Materials Science: Materials in Medicine*, ***20*(8)**, 1697-1708.

[11] Oporto G, Fuentes R, Álvarez H y Borie E **2008** Recuperación de la morfología y fisiología maxilo mandibular: biomateriales en regeneración ósea. *International Journal of Morphology*, ***26*(4)**, 853-859.

[12] Uvillús A **2018** Estudio de la deposición de hidroxiapatita sobre sustratos metálicos para aplicaciones biomédicas en presencia de campos magnéticos y por métodos electroquímicos (Bachelor's thesis, Quito-Ecuador).

[13] Chetty A, Wepener I, Marei M, Kamary Y y Moussa R **2012** Hydroxyapatite: synthesis, properties, and applications. *Nov Sci Publ*, 91-132.

[14] Laboratorio de Física de Materiales Dieléctricos y Ópticos Caracterización de materiales piezoeléctricos, [archivo pdf] recuperado el 13/11/2016 en http://www.ub.edu/gilafa/web_dielec/005/practica5.pdf

[15] Fernández F, Lozano, J y de Frutos Vaquerizo J **2003** *Introducción a la electrocerámica*. Fundación Rogelio Segovia para el Desarrollo de las Telecomunicaciones.

[16] Alves A **2010** Cerámicas piezoeléctricas: funcionamiento y propiedades Pereira. 1 São Carlos–Brasil. *Marzo de 1*.

[17] Fernández I, Alobera M, Canto D y Blanco L **2006** Bases fisiológicas de la regeneración ósea I: Histología y fisiología del tejido óseo. *Medicina Oral, Patología Oral y Cirugía Bucal (Internet)*, ***11*(1)**, 47-51.

[18] Baxter F, Bowen C, Turner I y Dent A **2010** Electrically active bioceramics: a review of interfacial responses. *Annals of biomedical engineering*, ***38*(6)**, 2079-2092.

[19] Montes F **2013**. Optimización de Materiales para regeneración ósea. Estudio de las interacciones de hidroxiapatita con proteínas y preparación de andamios porosos. Universidad Complutense de Madrid. (Doctoral dissertation).

[20] Londoño M, Echavarría, A y De La Calle, F **2006** Características cristaloquímicas de la hidroxiapatita sintética tratada a diferentes temperaturas. *Revista EIA*, **(5)**, 109-118.

[21] Dorozhkin S **2010** Bioceramics of calcium orthophosphates. *Biomaterials*, ***31*(7)**, 1465-1485.

[22] Tanaka Y, Kikuchi M, Tanaka K, Hashimoto K, Hojo J, Nakamura, M y Yamashita K 2010 Fast oxide ion conduction due to carbonate substitution in hydroxyapatite. *Journal of the American Ceramic Society*, ***93*(11)**, 3577-3579.

[23] Kumar D, Gittings J, Turner I, Bowen C, Bastida A y Cartmell S **2010** Polarization of hydroxyapatite: Influence on osteoblast cell proliferation. *Acta biomaterialia*, ***6*(4)**, 1549-1554.

[24] Caballero A, Fernández J, Duran P y Moure C 1994 Titanato de bario cerámico. *Boletín de la Sociedad española de cerámica y vidrio*, ***33*(1)**, 5-21.

[25] Perdomo C, Figueroa E y Páez J **2008** $Batio_3$ Obtenido Por El Método De Coprecipitación. *DYNA: revista de la Facultad de Minas. Universidad Nacional de Colombia. Sede Medellín*, ***75***, 223-230.

[26] Almeida A, Fechine P, Sasaki J, Ayala A, Góes J, Pontes D y Sombra A **2004** Optical and electrical properties of barium titanate-hydroxyapatite composite screen-printed thick films. *Solid state sciences*, ***6*(3)**, 267-278.

[27] Cadena J **2017** *Análisis y simulación del fenómeno de dispersión por modo de polarización (PMD) en fibras ópticas ITU-T G. 655* (Bachelor's thesis, Quito-Ecuador).

[28] Park J, Kelly B, Kenner, G. H., Von Recum, A. F., Grether, M. F., & Coffeen, W. W. **1981** Piezoelectric ceramic implants: in vivo results. *Journal of biomedical materials research*, ***15*(1)**, 103-110.

[29] Pereira A **2010** Cerámicas piezoeléctricas: funcionamiento y propiedades. ATCP Engenharia Física.

[30] Bowen C, Gittings J, Turner I, Baxter F y Chaudhuri J. 2006 Dielectric and piezoelectric properties of hydroxyapatite-$BaTiO_3$ composites. *Applied physics letters*, ***89*(13)**, 132906.

[31] Panda A **2011** *Preparation and characterization of Hydroxyapatite-Barium Titanate Composite* (Doctoral dissertation).

[32] Kokubo T y Takadama H 2006 How useful is SBF in predicting in vivo bone bioactivity?. *Biomaterials*, ***27*(15)**, 2907-2915.

[33] Castro M, Camargo J, Taub J, Fernandez J, Rubio F y Ramajo L **2013**. Cerámicos piezoeléctricos libres de plomo basados en los sistemas Bi0. 5 (Na0. 8K0. 2) 0.5 TiO3 y K0. 5Na0. 5NbO3 *Anales Acad. Nac. de Cs. Ex., Fís. y Nat.*, **65**, 50-58.

[34] Yamashita K, Oikawa N y Umegaki T **1996** Acceleration and deceleration of bone-like crystal growth on ceramic hydroxyapatite by electric poling. *Chemistry of materials*, **8(12)**, 2697-2700.

[35] Tang Y, Wu C, Wu Z, Hu L, Zhang W y Zhao K **2017** Fabrication and in vitro biological properties of piezoelectric bioceramics for bone regeneration. Scientific reports, **7(1)**, 1-12.

[36] Liu B, Chen L, Shao C, Zhang F, Zhou K, Cao Jy Zhang D **2016** Improved osteoblasts growth on osteomimetic hydroxyapatite/BaTiO3 composites with aligned lamellar porous structure. *Materials Science and Engineering: C*, ***61***, 8-14.

[37] Prakasam M, Albino M, Lebraud E, Maglione M, Elissalde C y Largeteau A **2017** Hydroxyapatite-barium titanate piezocomposites with enhanced electrical properties. *Journal of the American Ceramic Society*, ***100*(6)**, 2621-2631.

[38] Fang C, Hou R, Zhou K, Hua F, Cong Y, Zhang J y Cheng Y **2014** Surface functionalized barium sulfate nanoparticles: controlled in situ synthesis and application in bone cement. Journal of Materials Chemistry B, **2(9)**, 1264-1274.

[39] Mahabole M, Bahir M, Kalyankar N y Khairnar R **2012** Effect of incubation in simulated body fluid on dielectric and photoluminescence properties of nano-hydroxyapatite ceramic doped with strontium ions. Journal of Biomedical Science and Engineering, **5**, 396-405.

[40] Uribe R, Uvillús A, Bonilla O, Lascano L y González G **2021** Effect of Low Magnetic Field on Biomimetic Deposition of Hydroxyapatite on Titanium and BIOLINE Stainless Steel. Coatings, **11(12)**, 1484.

[41] Berzina L y Borodajenko N **2012** Research of calcium phosphates using Fourier transform infrared spectroscopy. Infrared spectroscopy-materials science, engineering and technology, 12(7), 251-263.

[42] Bhuiyan M, Alam M, Momin M, Uddin M y Islam M 2012 Synthesis and characterization of barium titanate (BaTiO3) nanoparticle. Int J Mater and Mec h Eng, **1**, 21-24.

[43] Vouilloz F, Castro M, Vargas G, Gorustovich A y Fanovich M **2017** Reactivity of $BaTiO_3$-$Ca10(PO_4)_6(OH)_2$ phases in composite materials for biomedical applications. Ceramics International, **43(5)**, 4212-4221.

[44] Dubey A, EA A, Balani K y Basu B **2013** Multifunctional Properties of Multistage Spark Plasma Sintered HA–$BaTiO_3$-Based Piezobiocomposites for Bone Replacement Applications. Journal of the American Ceramic Society, **96(12)**, 3753-3759.

Journal of Physics: Conference Series **2238** (2022) 012008 doi:10.1088/1742-6596/2238/1/012008

On the low star formation rate in Saggitarius C

Karen Martínez A.[1], **Ericson López**[2,3]

[1] Pontificia Universidad Católica de Valparaiso, Valparaiso, Chile
[2] Escuela Politécnica Nacional, Facultad de Ciencias, Departamento de Física, Quito, Ecuador
[3] Escuela Politécnica Nacional, Observatorio Astronómico de Quito, Quito, Ecuador

E-mail: karen.m91martinez@gmail.com

Abstract. The Central Molecular Zone (CMZ) contains approximately 10% of the total molecular gas in the Galaxy concentrated in turbulent and dense molecular clouds. The fact that the star formation rate is low in CMZ is an open question, since there is a large amount of dense gas clouds where star formation could have started. Sagittarius (Sgr) C is the only known star formation region located in the western side of the CMZ. The main goal of our study was to find out the dominant physical conditions of Sgr C and how they could affect the process of star formation in the CMZ. We studied the dust emission toward Sgr C using infrared data. Using radio data we also studied the kinematic temperature, star formation rate surface density and cloud morphology of Sgr C. Our study showed a dust temperature T_d of approximately 19 K toward Sgr C, which is in agreement with values estimated in other clouds of the CMZ and the Galactic plane. A high kinetic temperature higher than 58 K was determined toward Sgr C, which plays an important role in the cloud fragmentation and subsequent star formation. Based on our radio analysis we found a star formation rate surface density of $\Sigma_{SFR} = 4 \times 10^{-3}$ $M_{\odot} yr^{-1} kpc^{-2}$ which is lower than in others molecular clouds of the Galaxy. Finally, the HCO^+ emission reveals the presence of cavities and shells in regions where there is no evidence of star formation. These shells could be related to cloud shocks believed to take place in Sgr C. It is thought that cloud shocks may have triggered star formation in other sites of the Galaxy, which is not observed in Sgr C. A comparison between physical conditions in Sgr C with those of high star formation sources could allow to find the parameters that suppress the birth of stars in this region of the Galactic Center.

1. Introduction

The low star formation rate present in the Central Molecular Zone (CMZ) is an open question in the Galactic Center [1], due to the favorable conditions for such existence. Several studies have tried to answer this question, indicating that it could be due to the lack of dense high-mass nuclei or turbulence induced by cloud-cloud collisions [1]. Another approach based on gas dynamics, suggests that star formation is cycled and the CMZ is at a minimum of activity [2]. The study of dominant properties in the different molecular clouds of the Galactic Center could help to understand the reason for the low star formation rate in the CMZ.

Sagittarius (Sgr) C is a very characteristic region of the CMZ because this is the only region west of the CMZ that shows star formation that has been poorly characterized [3]. It has a size of 75×78 pc^2 and its center in galactic coordinates is located in $(-0.53, -0.107)$, approximately 75 pc from the Galactic Center. Average densities of 10^4 cm^{-3} [4] are found in this region, which makes Sgr C a high-density region. The most prominent structures present are a non-thermal filament (NTF), an HII region, and the only region with massive star formation, called EGO G359.44-0.102 [3]. The cloud with the highest activity in the complex is found at a speed of -65 km/s, this is where the star formation rate surface density was determined. The others physical properties like dust temperature and kinetic temperature of Sgr C were estimated in the center of Sgr C. Finally the qualitative study of the Sgr C morphology was carried out in the galactic coordinates $-0^{\circ}.7 \leq l \leq -0^{\circ}.2$ and $-0^{\circ}.3 \leq b \leq 0^{\circ}.12$ almost completely encompassing the molecular cloud complex Sgr C.

In each of the following sections, the different dominant physical properties in Sgr C were determined with the intention to compare the results with those obtained for other molecular clouds in the Galactic Center and trying to understand the causes of the low star formation rate present in this region of the CMZ. For this reason we divide this article in three sections. In section 2, the dust temperature (T_d) was estimated by a far-infrared (FIR) study and the kinetic temperature (T_k) by a radio study. In section 3 the star formation rate surface density (Σ_{SFR}) was estimated using radio data. In section 4 a qualitative morphological study of dense gas movement was carried out in the entire molecular complex Sgr C. Finally, in sections 5 and 6 the general discussion and the conclusions obtained in this research project are shown.

2. Sgr C physical properties

To estimate the physical properties of dust temperature T_d and kinetic temperature T_k in Sgr C we work at far-infrared (FIR) and radio wavelengths respectively.

2.1. Dust temperature

Interstellar dust catalyzes the chemical reactions that occur in the formation of molecular hydrogen and other molecules that characterize molecular clouds [5]. Dust also protects the interior of clouds absorbing high-energy photons and cosmic rays. It also controls the heating and cooling processes in the densest regions of molecular clouds [5]. When dust absorbs radiation from starlight, the dust grains can release photo-electrons which, by absorbing optical and ultraviolet (UV) light, heat the dust grains and then radiate in the FIR where can be detected [6]. Therefore, that the dust emission is thermal, i.e., could be modeled by a Black Body. In addition the thermal radiation of dust depends on the temperature of the study region [5], then at FIR wavelengths is used to estimate the T_d, present in the central regions of Sgr C. To perform this estimation we used the following gray body equation modified by the body emissivity (β). Thus, the Planck function (B_λ) is used as a function of wavelength according to eq. (1) [7]:

$$B_\lambda(T_d) \approx \frac{2\pi hc^2}{\lambda^5} \frac{\lambda^\beta}{e^{\frac{hc}{\lambda k T_d}} - 1} \tag{1}$$

2.1.1. Data For the photometric study, FIR dust emission maps were obtained from the *Herschel Space Telescope* using the *Herschel Science Archive database (HSA)* [8]. The Herschel Space Observatory focused on photometry and spectroscopy in the FIR and the submillimeter range covering the 60-670 μm band. Herschel's objective was to study the cold universe with temperatures between 5-50 K [8]. In this project, a set of 5 dust emission maps obtained with the PACS and SPIRE instruments was used, for which the HSA database [8]. To study the region located in the CMZ, for this study the PACS and SPIRE photometric maps were used. PACS and SPIRE allow tracking the peak spectral energy distribution (SED) of cold dust and thus the temperatures and luminosities of the ISM star-forming complexes [9].

2.1.2. Determination of Spectral Energy Distribution: To estimate the T_d, it was necessary to determine the SED that was later fitted using Planck's law shown in eq. (1). First, the value of the flux density (Jy) was determined at different wavelengths (μm) by aperture photometry using *HIPE*, it is a map processing system, with an aperture radii of 130 arcsec. Then the values obtained with their respective errors were fitted, thus obtaining an estimate for the dust temperature. The values obtained for energy flux at each wavelength are shown in Table 1.

Table 1. Flux density values and errors from the photometric analysis of the center of Sgr C.

Wavelength (μm)	Flux density in Sgr C (Jy)	Associated error in Sgr C (Jy)
70	25247.17	1262.36
160	27634.51	1381.73
250	11356.57	567.83
350	4507.58	225.38
500	1495.68	74.78

2.2. Kinetic temperature

The estimate of the T_k is done based on the excitation temperature (T_{ex}), which is the temperature that relates two energy levels in a Boltzmann distribution (since the excitation of the molecules depends on the collisions) under conditions of local thermodynamic equilibrium (LTE), where the velocity of the particles are easily thermalized and as the energy levels are dominated by collisions, the T_k can be approximated by T_{ex} ($T_k \sim T_{ex}$)[10]. The T_{ex} can be obtained from spectroscopy using a rotation diagram through the use of the MADCUBA program, that resolve the expression (2) [12]:

$$\log\left(\frac{8\pi k\nu^2}{hc^3 g_u}\int T_L dv\right) = \log\left(\frac{N}{Q(T_{ex})}\right) - \frac{E_u}{kT_{ex}}\log(e) \qquad (2)$$

where $h = 6.626070 \times 10^{-34}$ Js is the Planck constant, $k = 1.3806485 \times 10^{-23}$ J/K is the Boltzmann constant, ν is the observation frequency in s^{-1}, T_L is the line intensity in temperature function in K, N is the column density cm^{-2}, $Q(T_{ex})$ is the temperature dependent partition function of the molecule, g_u is the statistical weight of level u and E_u is its energy above the ground state [12]. Therefore, knowing the temperature of the line $T_{L,\nu}$ obtained from the spectral emission of the dense gas tracer, the excitation temperature can be obtained by varying each of the parameters of the eq. (2) [12].

The Gaussian fit determined parameters present in data from molecular tracers like $FWHM$ and V_{LSR}, in order to obtain information on the physical properties of the tracers. For this, studies of the Gaussian profile of the spectra, such as speed and dispersion, are carried out. This facilitates the determination of the physical properties of the gas in a region. From the relation (2) we obtain the Gaussian fit law [12]:

$$\tau_v = \tau_0 e^{\frac{-4\ln 2(v-v_0)^2}{\Delta v^2}} \qquad (3)$$

where $\Delta v = 2\sqrt{(2\ln 2)}$, $\sigma = 2.354\ \sqrt{(kT_k/m)}$ is the Full Width at Half Maximum (FWHM) of the Gaussian profile and τ_ν is the opacity at the velocity ν, and τ_0 is the opacity at the initial velocity ν_0 [12].

Journal of Physics: Conference Series **2238** (2022) 012008 doi:10.1088/1742-6596/2238/1/012008

2.2.1. Data To estimate the T_{ex} in the emission of acetonitrile molecule (CH_3CN), we use the data cube of CH_3CN in the CMZ obtained from the database of the Australian Telescope National Facility (ATNF) of the *MOPRA* radio telescope, the GILDAS [11] processor was used for the data cube processing, while the MADCUBA [12] processor was used for the Gaussian fit.

The *MOPRA* telescope is used for spectroscopy in various frequency ranges at millimeter wavelengths. For this research project we worked in the 3 mm band that covers a frequency range from 76 to 117 GHz [7].

2.2.2. Determination of excitation temperature: For the derivation of the kinetic temperature T_k, the T_{ex} of the molecular tracer CH_3CN was first estimated. In this project, T_{ex} value is equivalent to the T_k under the conditions of LTE [10]. To determine the T_{ex}, an averaged spectrum of the study region was obtained, which was subsequently fitted by means of a Gaussian modeling of synthetic spectra with which an estimate of the T_{ex} and therefore the T_k of the region was obtained.

The speed of the source was set at −53 km/s for the study in Sgr C obtained from Doppler effect analysis with GILDAS, thus a value of 10 km/s for the line width of Sgr C was obtained from the Gaussian fit model with synthetic spectra.

2.3. Results and discussion

The results obtained for the kinematic study are shown below in both the T_d and the T_k.

The estimation of the dust temperature in the center of Sgr C is shown in Figure 1.

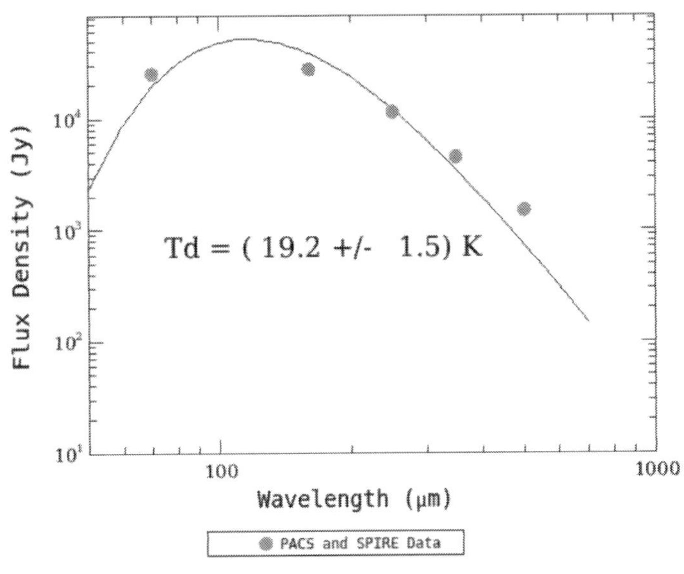

Figure 1. Spectral distribution of energy in the center of Sgr C. Fluxes values at five different wavelengths are shown in blue.

Figure 1 shows the SED, for the region of the center of Sgr C. This graph shows the errors associated with the values of flux density, these values are small compared to the estimated values for fluxes in around 5%, so it does not have a noticeable effect on the representation of SED. Here it can be seen that the flux density values obtained for the center of Sgr C have the presence of a maximum of thermal dust emission at a wavelength of 160 μm, this peak corresponds to maximum emission for dust in the FIR range. For the Center of Sgr C, the gray body fit curve with β fixed in 1.5, roughly describes the values obtained by the SED, with a very marked difference at the wavelength of 160 μm where the value of the flux density is much lower. The estimated value for the dust temperature is $T_d = 19.2 \pm 1.5$ K is reliable for this region of the molecular complex Sgr C. From the best fit model, it was possible to identify that the wavelength of a maximum emission in the FIR of the dust is approximately equal to 120 μm, which corresponds to a temperature of approximately 20 K, as reported in the literature for the Galactic Center clouds [9].

The values obtained for the estimation of the kinetic temperature in the study region Sgr C was (58.1 ± 8.4) K. This values and other are shown in Table 2.

Table 2. Results of the kinematic study of the molecular complex Sgr C. Parameters obtained from the Gaussian fit and estimation of the T_k with their respective error.

Source	Gal. Coord. [l,b]	FWHM [km/s]	V_{LSR} [km/s]	T_k [K]	Error T_k [K]
Sgr C	-0.57,-0.10	10	-53	58.1	±8.4

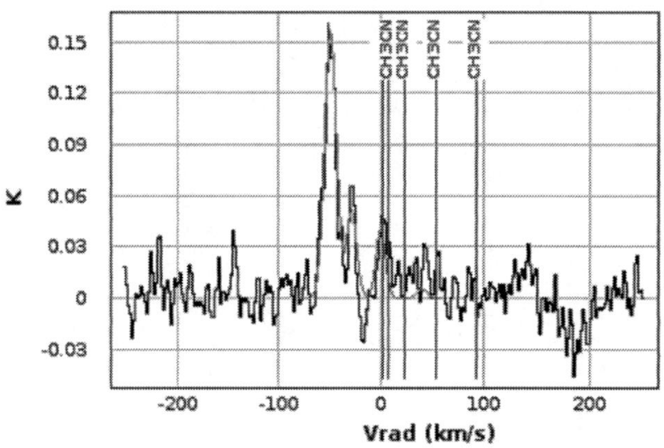

Figure 2. Result of the Gaussian fit to estimate the T_k in the center of Sgr C. The original average spectrum is shown in black, the Gaussian fit by synthetic spectra in red and the transition lines position or rest frame lines of CH_3CN in blue.

In Figure 2, the spectrum for the center of Sgr C is shown. The average spectrum of the molecular emission of acetonitrile in the center of Sgr C is shown in black, the profiles of the synthetic spectrum in red and the transition lines of CH_3CN in blue. It shows that the fitted made by the MADCUBA software is reliable. Also it is observed that the line profiles for Sgr C are narrow. This and the fact of the presence of an EGO G359.44-0.102 near of this position indicate that it is possible that the relatively low temperature obtained T_k of approximately 58 K is because the effects of the EGO are not being shown in the environment due to its evolutionary state.

3. Stellar Rate Formation

The massive star formation possesses energetic stellar winds and supernova blast waves that drive the dynamical evolution and morphology of the ISM. They can regulate the star formation rate, the galaxies evolution and are responsible for ionizing the interstellar medium, which contributes through this feedback to the reionization of the universe [14]. Despite these important features, a little is known about how high-mass star formation occurs because they are rare, although some aspects can be explained by the initial mass function (IMF), and their formation sites are far away [14]. In addition, the zero main sequence of massive star and their formation process is shorter; and in the Galaxy it occurs deep within the densest parts of molecular clouds at the Galactic Center [15].

The star formation rate (SFR) in a molecular cloud is directly controlled by the amount of dense molecular gas present inside [13]. To establish the relationships necessary to express this proportionality, the Kennicutt-Schmidt (K-S) relationship is used, which expresses that stars are formed exclusively in regions of molecular gas [16]. To study the correlation that exists, a power law is assumed as shown in eq. (4).

$$\Sigma_{SFR} \propto \Sigma_{gas}^{N} \tag{4}$$

where N, is the power law index, Σ_{SFR} is the star formation rate surface density that is defined as the speed with which stars are formed per unit area and Σ_{H_2} is the molecular hydrogen surface density. In the case of the Milky Way, the Kennicutt-Schmidt relationship is expressed in eq. (5) as follows [16]:

$$\Sigma_{SFR} = 10^{-2.1\pm0.2}\left(\frac{\Sigma_{H_2}}{10M_{\odot}pc^{-2}}\right)^{1.0\pm0.2} (M_{\odot}yr^{-1}kpc^{-2}) \tag{5}$$

To determine the molecular hydrogen surface density (Σ_{H_2}), the intensities in the emission lines of the carbon monoxide ^{13}CO isotope are measured by its optically thin lines because it is stable and it is the second most abundant molecule after molecular hydrogen and to which their column densities are related. Eq. (6) is used to estimate (Σ_{H_2}) from the carbon monoxide isotope [16].

$$\Sigma_{H_2(^{13}CO)} \approx 10^{-33} \times m_H \times \left(\frac{N_{H_2(^{13}CO)}}{cm^{-2}}\right) \times \left(\frac{A_{pixel}}{cm^2}\right) \times \left(\frac{A_{cloud}}{pc^2}\right)^{-1} (M_{\odot}pc^{-2}) \tag{6}$$

where $m_H = 1.67 \times 10^{-24}$g is the mass of hydrogen, A_{cloud} is the area of the cloud expressed in pc^2 y A_{pixel} is the area of the pixel given in cm^2. And finally, with the results of eq. (6) the star formation rate surface density was determined by eq. (5).

3.1. Data

To determine observational parameters, such as the cloud area, the pixel area and the column density of ^{13}CO, the tracer data cube was needed. The cube was obtained by means of the *Nobeyama* radio telescope (NRO) which is a branch of the National Astronomical Observatory of Japan (NAOJ). This is the world's largest single-dish radio telescope in the 100 gigahertz band and specializes in approximately 3-millimeter spectroscopy with a high sensitivity and angular resolution of 17 ± 1 arcsec in a frequency range of [$\sim$ 85, 116] GHz. The data cube of carbon monoxide ^{13}CO in the transition J=1-0 was obtained at a frequency of 110.20 GHz in the CMZ [17].

3.2. Determination of star formation rate surface density

To determine the star formation rate surface density and plot the Kennicutt-Schmidt relationship, eq. (5) was used. For this, the Σ_{H_2} was estimated by means of the eq. (6). To use this equation, the column density of H_2, the area of the cloud and the area of a pixel of the study region had to be determined by spectroscopic analysis with GILDAS; these last two parameters were obtained by spectroscopy processing of an integrated velocity map between [-80, 60] km/s for the data cube of the molecular emission of carbon monoxide ^{13}CO.

Journal of Physics: Conference Series **2238** (2022) 012008 doi:10.1088/1742-6596/2238/1/012008

3.3. Results and Discussion

As results for the study of the star formation rate in the molecular cloud complex Sgr C, the area of the cloud, the molecular hydrogen surface density and the star formation rate surface density were estimated. The logarithmic values of these last two quantities where calculated to be able to graph the star formation rate surface density in the K-S relationship plot to compare it with other galactic clouds. In this case, it was assumed that the surface density of molecular hydrogen is equivalent to the surface density of the gas since, being a molecular cloud, the largest amount of material present is precisely molecular hydrogen.

Table 3. Results of the study of star formation in the molecular complex Sgr C from the molecular tracer ^{13}CO. In the estimated parameters Σ_{SFR} and Σ_{H_2}, the natural logarithm was calculated to be able to locate it in the K-S relationship.

Star formation property in the Sgr C	Result	Units
cloud area	392.62	[pc^2]
Σ_{H_2}	8.75	[M$_\odot pc^{-2}$]
Σ_{SFR}	0.004	[M$_\odot yr^{-1} kpc^{-2}$]
$\log(\Sigma_{H_2}[M_\odot pc^{-2}])$	0.9419	-
$\log(\Sigma_{SFR}[M_\odot yr^{-1} kpc^{-2}])$	-2.37	-

Table 3 shows the results of the study of the star formation rate in the molecular complex Sgr C. The values of $\log(\Sigma_{H_2})$ and $\log(\Sigma_{SFR})$ were used to locate the molecular hydrogen surface density (which corresponds to the star-forming gas in the ISM) and the star formation rate surface density in the Schmidt-Kennicutt diagram that relates $\log(\Sigma_{SFR})$ vs. $\log(\Sigma_{H_2})$, as shown in Figure 3.

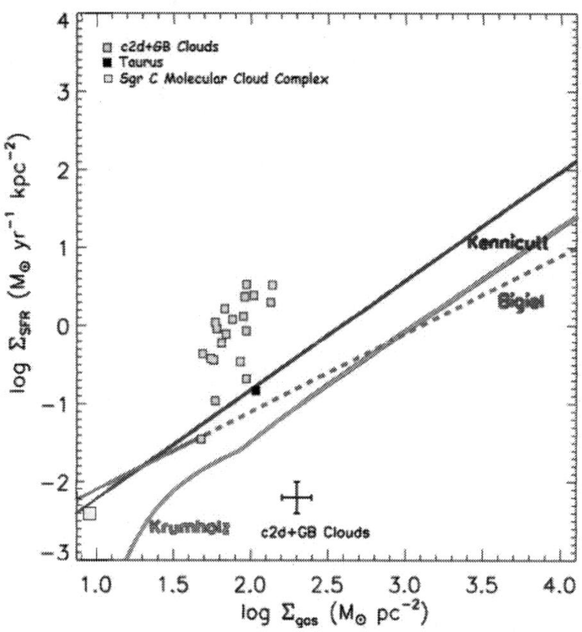

Figure 3. Result of the estimation of the Σ_{SFR}: K-S relationship for the molecular complex Sgr C.

Figure 3 shows the Kennicutt-Schmidt relationship that relates the star formation rate surface density and the molecular hydrogen surface density. The base plot shows a study of galactic star formation regions performed on a total of twenty molecular clouds along the Galactic Disc (cyan square markers) [16]. The result obtained in the star formation study for the molecular complex Sgr C is shown in a yellow square. The different lines indicate the various laws that predict star formation in the Milky Way [16]. In the case of the star formation of the molecular complex Sgr C, it keeps the necessary conditions to present star formation like high density about $10^4 cm^{-3}$ [4], dust temperature of approximately 19 K and kinetic temperature of 58 K. However, we quantitatively confirmed that it presents a star formation rate surface density that is too low compared to other clouds and molecular complexes, which according to the literature have higher stellar formation in much smaller areas as the Perseus complex of 1.32 $[M_{\odot} yr^{-1} kpc^{-2}]$ in an area of 73.2 pc^2 or the Taurus cloud complex with 0.147 $[M_{\odot} yr^{-1} kpc^{-2}]$ in an area of 252 pc^2 [16]. Therefore there must be conditions that are inhibiting the formation of stars in this region as well as the dynamics of the morphological structures in this molecular complex [1]. Other causes that explain this low star formation in this region so characteristic of the CMZ could be a modification in the star formation processes [2].

4. Morphological study in the Sagittarius C Molecular Complex

Star formation is linked in a complex way to the gas structures present in molecular clouds and the dynamics that characterize the internal structure due to the effects of turbulence, gravity and magnetic fields [18]. Every structure present in molecular clouds, such as bubbles, layers, filaments, rings and cavities vary in shape from almost spherical to irregular and can interact and influence star formation, because it is a highly dynamic process that modifies the physical environment of the clouds [18]. These structures have been detected and studied in wide ranges of the electromagnetic spectrum [19].

A qualitative characterization of morphological structures allows to determine the physical mechanisms of where and how many stars can form in molecular clouds by means of studies in radio with molecular tracers; however, each molecular species is sensitive only in the range of density where the information was obtained. It is important to mention that this can get biased depending on the molecule used [20]. For this reason in this research three tracers (dust thermal, formyl cation and atomic hydrogen) at different wavelengths (350 μm, 89.19 GHz and 1.42 GHz) were used to study the structures present in the molecular complex Sgr C. The evolution of baryonic matter in galaxies is driven mainly by the formation of stars that could be closely related to the orbital dynamics of the gas flow in the molecular clouds of the CMZ. To study the gas flow in three dimensions of the CMZ there are three scenarios which are described below [20]:

- **Spiral arms:** It shows a good qualitative approximation of the kinematics of the gas and suggests that it is distributed in two spiral arms that create a ring structure at approximately 120 pc, where the arm closest to the GC harbors Sgr B (there is high star formation in this complex) and the most distant arm harbors Sgr C.
- **Elliptical closed orbit:** Here, the gas flow shows an open elliptical ring of gas and dust due to the extended mass distribution in the CMZ and where there is a temporal evolution of the clouds for the abundance of dust between the Brick and Sgr cloud B. In this case the gas is shaped like a "crooked ring".
- **Open flow:** This case was developed based on a dynamic model of the CMZ. Here, the orbital dynamics are divided into four open flows and there is a time line for the dust crest clouds in the region of active star formation in the eastern part of the GC that is consistent with the observations of clouds and molecular complexes in the area.

4.1. Data

To determine the composition of the maps in the three molecular emissions GILDAS software was used and three sets of public access data were obtained from three telescopes, two on radio frequencies: *ATNF* (Australia Telescope National Facility) [7], *NRAO* (National Radio Astronomy Observatory) [17] and one on infrared wavelengths: Herschel (Herschel Spacial Telescope) [8]. To make the maps of the spatial distribution with the dense gas tracer, the data obtained from the *MOPRA* radio telescope was used, where the formyl cation HCO^+ observed at a resting frequency of 89.19 GHz was used as a molecule. HCO^+ molecular emission data were obtained from the ATNF telescope database. The overlay was carried out using the integrated maps of the HCO^+ tracer. The dust emission as a base in infrared, at a wavelength of 350 μm, was obtained from the HSA. And the radio-continuous absorption of the HI at 1.42 GHz was obtained from the VLA radio telescope database of the National Radio Astronomy Observatory (NRAO) [21].

4.2. Dense gas spatial distribution maps

To study qualitatively the morphology of the structures present in the molecular complex Sgr C, three tracers were used that allowed us to observe different regions of the ISM: the infrared emission of the dust, the radio molecular emission of the formyl ion and the absorption in radio-continuum of neutral hydrogen. With the HCO^+ tracer, a spatial distribution of the dense gas was made in different speed ranges as contours, this in conjunction with a superposition of the emissions from dust (at a wavelength of 350 μm) and the emission in radio-continuum of the HI allowed the acquisition and interpretation of detailed maps of the morphological structure of the Sgr C complex.

4.3. Results and Discussion

From the study of morphological structures in the molecular complex Sgr C, a composition of nine maps is shown in Figure 4. They were obtained from the superposition of the nine speed ranges of the emission contours of the HCO^+, shown in green; the infrared dust emission at a wavelength of 350 μm is shown in gray scale; and the map of the absorption contours of molecular hydrogen is shown in lavender.

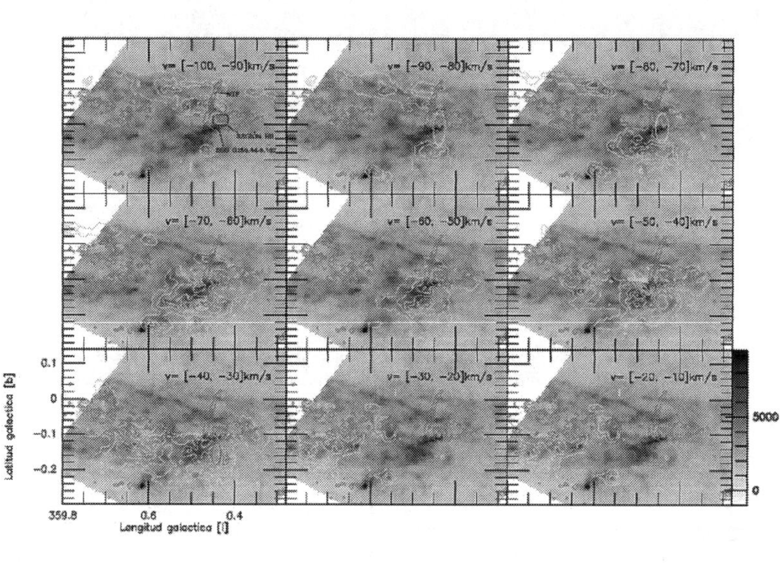

Figure 4. Result of the qualitative morphological study of the molecular complex Sgr C.

These maps show the characteristic filament at position (−0.55, −0.09) in galactic coordinates, at the base of which is the EGO G359.44-0.102. Also, a cavity in which star formation could be hosted is shown in yellow in the second and third panel in Figure 4. The reference points are the center of Sgr C (cyan point) and the EGO G359.44-0.102 (blue point) that helped us to verify how the dense gas formyl ion moves along the molecular complex Sgr C in relation to the reference point Sgr C. In Figure 4, we can observe that the cloud moves from north to south until reaching a speed interval of [−70, −40] km/s associated with the molecular complex Sgr C in panels 3, 4 and 5. Here, we observed the presence of cavities near the HII region, under the star formation zone G359.44-0.102, where there is no emission of dense gas in the velocity range of [−100, −60] km/s. After that we could observed that the emission of dense gas moves again from south to north in the Galactic Center direction in the speed range of [−40, −20] km/s where the gas emission is already due to sources other than those found in Sgr C. In the speed range [−50, −40] km/s we can observe the maximum gas emission associated with star formation in the EGO.

In this qualitative study of the morphological structures present in the molecular complex Sgr C, we took into account that the emissions made it possible to trace the HII regions by the dust emission and the gas present in the molecular complex Sgr C. For different tracers it was possible to see that the gas is related to the HII regions; we could find the presence of a circular cavity under the HII region, where there is the presence of the EGO G359.440.102, in the speed range [−100, −70] km/s. This cavity could be the cause of star formation inhibition in this region, because it is a clear evidence of the presence of strong stellar winds, which dissipate gas agglomeration in molecular clouds for the motion of dense gas due to the presence of ionized gas sheets that exist around the HII region [22].

5. Conclusions

As Sgr C is a poorly characterized region in the western CMZ that exhibits massive star formation, determining its dominant physical properties was relevant to try to understand the low rate of star formation in this region. According to this approach, the following conclusions are presented:

- The values obtained in this project of 19 K for dust temperature and 58 K for kinetic temperature in the molecular complex Sgr C indicate that this is a region located in a interstellar medium with high densities in its interior. For these reasons, the molecular complex Sgr C complies with favorable characteristics present in well-known molecular clouds that show massive and active star formation within.
- The star formation rate surface density estimated in the molecular complex Sgr C is 0.004 $M_{\odot} yr^{-1} kpc^{-2}$ for an area of 392.62 pc^2. It quantitatively confirms that the molecular complex Sgr C has a practically inactive star formation rate surface density compared to its Galactic homologous [16], which show high star formation per unit area and time despite having relatively small areas showing continuous star formation, compared to that used in this research.
- Based on a qualitative study of the morphological structures and the motion of the dense gas HCO^+ in the molecular complex Sgr C, which is an excellent tracer of cold gas nuclei of high molecular hydrogen density [23], we concluded that the molecular gas cavity found, under the HII region where the EGO G359.44-0.102 is present, in the speed range [−100, −70] km/s, is consistent with the study carried out by Lang [21]. This cavity could be the cause of massive star formation inhibition in this region of the Sgr C molecular complex, as it is clear evidence of the presence of strong stellar winds, as indicated in the work by Liszt and Spiker (1995) [24]. These winds dissipate gas agglomeration in molecular clouds due to the presence of ionized gas layers that exist around the HII region studied by Lang [21].
- The characterization of physical properties in the only region that shows evidence of star formation in the western part of the Central Molecular Zone allowed us to establish that the

physical conditions of density and temperature present in this region could not be the cause of a low rate of star formation in this CMZ region since the estimated values are consistent with studies carried out in molecular clouds that present active star formation. Some of the causes that could be slowing down the rapid production of stars in this region of the CMZ could be the absence of dense cores, a different mechanism of star formation or the effects of gas dynamics in the region as analyzed qualitatively in this research. Therefore, the systematic and comparative observational studies of the estimated physical properties in the molecular complex Sgr C and the relationship of these properties with the different structures present in the molecular clouds allowed us to characterize the star formation inside the molecular complex Sgr C in the part internal of the CMZ.

6. Acknowledgments

The authors express their gratitude to Dr. Jairo Armijos, researcher at the Astronomical Observatory, who proposed this interesting topic of study and gave the necessary guidelines for carrying out this work.

References

[1] Kauffmann J 2015 *Central Molecular Zone of the Milky Way: Star Formation in an extreme Environment* vol 11, (Proceedings of the International Astronomical Union) p 163-166
[2] Kruijssen D 2016 a multi-scale understanding of the gas-star formation cycle in the Central Molecular Zone vol 322, (The Multi-Messenger Astrophysics of the Galactic Centre, Proceedings of the International Astronomical Union, IAU Symposium) p 64-74
[3] Kendrew S and Ginsburg A 2013 *Early-stage Massive Star Formation near the Galactic Center: Sgr C* vol 775, (The Astrophysical Journal Letters) p L50
[4] Tanaka K and Oka T 2014 *High Velocity Compact Clouds in the Sagittarius C Region* vol 783, (The Astrophysical Journal) p 62
[5] Chira R 2018 *On Filaments within Molecular Clouds and their Connection to Star Formation* (arXiv e-prints)
[6] Williams D A and Viti S 2010 *Observational Molecular Astronomy: Exploring the Universe using Molecular Line Emissions* (UK: Cambridge University Press)
[7] Wong G F 2016 *Characterising the Structure of Molecular Clouds* (Sydney: Western Sydney University)
[8] Molinari S and Swinyard B 2010 *Hi-GAL: The Herschel Infrared Galactic Plane Survey* vol 122, (Chicago: Publications of the Astronomical Society of the Pacific) p 314
[9] Romero R 2008 *Población estelar joven embebida en la nube molecular galáctica asociada a la fuente IRAS 18236-1205* (Puebla: Instituto Nacional de Astrofísica, Óptica y Electrónica)
[10] Silva A 2010 *Analysis and Simulation of Semiconductor Devices* (Chile: UNIVERSIDAD DE CHILE)
[11] Gildas Team 2013. *GILDAS: Grenoble Image and Line Data Analysis Software*
[12] Martín S and Martín-Pintado J 2019 *Spectral Line Identification and Modelling (SLIM) in the MAdrid Data CUBe Analysis (MADCUBA) package. Interactive software for data cube analysis* vol 631, (Astronomy and Astrophysics) p A159
[13] Lada C., Forbrich J. 2012 *Star Formation Rates in Molecular Clouds and the Nature of the Extragalactic Scaling Relations.*, p 190,
[14] Lada C and Pudritz R 1988 *Galactic and Extragalactic Star Formation* (Springer Netherlands)
[15] Armijos J 2014 *Química en las nubes moleculares del Centro Galáctico* (Madrid: UAM. Departamento de Física Teórica; CSIC. Centro de Astrobiología)
[16] Heiderman A and Evans N 2010 *The Star Formation Rate and Gas Surface Density Relation in the Milky Way: Implications for Extragalactic Studies* vol 723, (The Astrophysical Journal) p 1019-1037
[17] Oka T and Hasegawa T 1998 *A Large-Scale CO Survey of the Galactic Center* vol 118, (The Astrophysical Journal Supplement Series) p 455-514
[18] Froebrich D and Rowles J 2010 *The structure of molecular clouds - II. Column density and mass distributions* vol 406, (Monthly Notices of the Royal Astronomical Society) p 1350-1357
[19] Gahm G F 1994 *Interstellar cloud morphologies and their origins* vol 3, (Baltic: Baltic Astronomy) p 85-103
[20] Hacar G A 2013 *Formación de núcleos densos en las nubes moleculares de Tauro y Perseo* (Madrid: Universidad complutense de Madrid)

[21] Lang C C and Goss W M 2010 *A High-resolution Survey of HI Absorption Toward the Central 200 pc of the Galactic Center* vol 191, (The Astrophysical Journal Supplement Series) p 275-300
[22] Beuther H and Churchwell E B 2007 *The Formation of Massive Stars* ed Reipurth B and Jewitt D (Protostars and Planets V) p 165
[23] Williams D A and Hartquist T W 2013 *The Cosmic-Chemical Bond* (The Royal Society of Chemistry)
[24] Liszt H S and Spiker R W 1995 *Radio Images of Sagittarius. II. Sagittarius C and Its Neighbors* vol 98, (The Astrophysical Journal Supplement Series) p 259

JWST retrieval simulation of super-Earth

H Barbier[1], E D Lopez Izurieta[1,2], A Manzaba Vélez[1] and D A Herrería Hoyos[1]

[1] Escuela Politécnica Nacional, Facultad de Ciencias, Departamento de Física, Quito, Ecuador
[2] Escuela Politécnica Nacional, Observatorio Astronómico de Quito, Quito, Ecuador

E-mail: hugo.barbier@epn.edu.ec

Abstract.
The James Webb Telescope (JWST), launched at the end of 2021, will permit the study of exoplanets' atmospheres with an accuracy never obtained before. One of the most keenly anticipated science themes with JSWT data is the study of super-Earth type planets. There exist many questions about their interior and atmospheric composition as well as their evolution and possible migrations.

In this work, we simulated transmission spectra of super-Earth type systems with the petitRADTRANS code and emission spectra of a M-type star with PySynphot. The primary transit observation is simulated with the MIRISim simulator working in the MIRI LRS-Slitless mode. We then obtained the transmission spectra of the planet using a code that we developed ourselves. Finally, we carried out the retrieval: first, in order to check the retrieval tool included in the petitRADTRANS code, we treated the original transmission spectrum. Second, the code is used with the spectrum obtained from the telescope simulator.

The results found encourage us to compare them to results found when using other tools, with the hope of gaining precision in those results, in addition to working in a wider range of wavelengths.

1. Introduction

The JWST has several scientific missions. One of them is the detection, observation and characterization of exoplanets; particularly the study of their atmospheres. Its instrumentation must be understood to find the most useful modes for characterizing exoplanets [3, 10, 26]. This is a hard task since there are hundreds of possible combinations of modes and filters. Thus, it is important to simulate data that the JWST could send us during its observations. Particularly, the measurement limitations that we have for the detection and characterization of exoplanets are important to know in order to study the instrument performance, optical and geometric effects as well as the creation of realistic spectra of exoplanets and their atmospheres.

In this work we use the MIRISim simulator [9], developed by the European consortium of MIRI [6], and the petitRADTRANS (pRT) code developed by Molliere [24]. We focus on the MIRI instrument operating in the LRS-Slitless mode which works between 5 and 12 micrometers. This mode is known to be more suitable for the study of exoplanets transit spectroscopy [2]. The LRS-Slitless mode of the MIRI instrument allows us to study the spectra of exoplanets, which in our case is the super-Earth transmission spectra around red dwarfs. The spectra can allow us to partially probe the composition of the planets' atmospheres and structures, as well as its evolutionary history and possible habitability. There are computational tools available which are in continuous development, which helped us to simulate the data, create observation scenes and choose the correct instruments for observation. Some objectives of this work are understand the different tools available for the simulation and retrieval of spectra, their sensitivity to abundance of some molecules such as H_2O and CH_4, and identify the problems that may arise in future work. These simulations play an important role in learning methods that can be applied to reduce the exoplanet data to be observed with JWST.

This manuscript has been organized as follows: Section 2 gives an introduction to the super-Earth planets which are the targets of our study. Section 3 briefly reviews the JWST characteristics, particularly focusing on components that aid in studying exoplanets. In Section 3, we present the MIRI instrument and the LRS-Slitless mode. Section 4 describes simulators we have used and explains the MIRISim simulator. Also, the PySynphot is introduced. In Section 5, the target (star and super-Earth planet) simulated in this work is presented. The results and the data analysis are summarized in Section 6. Section 7 is dedicated to the final discussion and conclusions.

2. Super-Earths

Among the large diversity of exoplanets discovered by the Kepler mission, super-Earths (and mini-Neptunes) are the most abundant, with masses between 1 and 10 Earth mass [4]. These planets, with intermediate mass in the range between terrestrial planets and gas giants, do not exist in our solar system and the fundamental questions about their origin, structure, and composition are still open [21] [20]. These kind of planets are both larger than Earth and more massive. Their surfaces are likely surrounded by a gas envelope. Planets with 1.75 Earth radius are massive enough to have a thick atmosphere and a substantial portion of hydrogen-helium (H/He) [20] envelopes that are bound through gravity, but not so much like Neptune or mini-Neptune planets [17]. Thus, we choose to study the upper size limit for super-Earths in this work as target for our simulation to learn about the atmospheric composition.

As a super-Earth has intermediate size, the periodic flux decreases during the occultation of a M-type dwarf companion by about 0.1 to 1%, much more than the decrease in flux when the Earth orbits the Sun (10^{-4} %) [8] [20]. This relatively large decrease in flux is one of the reasons for choosing these targets for our simulation.

3. JWST Mid-InfraRed Instrument (MIRI) and LRS-Slitless Mode

The JWST is a well equipped telescope with four important instruments on board: the Near-Infrared Camera, or NIRCam [3], the Near-Infrared Spectrograph, or NIRSpec [12], the Mid-Infrared Instrument, or MIRI [6], and the Near InfraRed Imager and Slitless Spectrograph or NIRISS [15]. Each of these instruments operates in several observation modes: imaging and low, medium or high resolution spectroscopy, with or without the use of coronagraphs and filters. The sensitivity of the instruments is over a wide wavelength range from the optical to the mid-infrared (0.6 - 28.3 μm) corresponding to black-body emission peaks of temperature between 6000 K and 100 K, capable of covering a wide range of molecular resonance frequencies [15].

One important instrument of the JWST is the Mid-InfraRed Instrument (MIRI), which provides imaging and spectroscopic observing modes. This instrument operates in the range between 4.9 to 28.8 μm, it has a wide-field broadband camera, a Medium Resolution Spectrograph (MRS) and a Low Resolution Spectrograph (Slit or Slitless modes working between 5 and 12μm) that allows observations of bright targets. The MIRI LRS-slitless mode uses the sub-array that allows the dynamic range of the LRS to be increased, up to a readout time of about 0.159 s; this upper limit is important to avoid saturation during the observation process. The wide field slitless spectroscopy is adequate for a dedicated target time-series spectroscopy or spectroscopic studies during transits. High precision spectrophotometry over long observations is useful for characterization of transiting exoplanets, for that the MIRI must operate the LRS

Journal of Physics: Conference Series **2238** (2022) 012009 doi:10.1088/1742-6596/2238/1/012009

in a slitless mode. In general, it is also ideal for the observation of bright point-sources. For faint targets, the slit is necessary in order to obtain a better performance. However, the signal to noise ratio is reduced in the slitless mode because the background is higher due to the absence of the slit. The sensitivity is also affected, it is 10 times lower than in the LRS slit mode (see JWST noise documentation).

4. Simulations tools

4.1. JWST-MIRI Instrument: MIRISim simulator

We have simulated the exoplanets' observation using the MIRISim that is a simulator still under development, but already available to researchers for simulations [9]. It is the official MIRI simulator that models the behavior of this instrument and its data products. Other teams have already tried to simulate data in different ways [16, 18, 13]. It is able to model Imager, LRS and MRS simulations and the photon propagation through MIRI. This Python-3 tool has been developed by the MIRISim team, part of the MIRI European Consortium.

Each time the MIRISim simulator runs, it uses calibration files that are downloaded automatically to simulate the different instrument effects (linearity, saturation, distortion, flat-field, fringing, etc). It uses the information and data collected during the cryogenic test and calibration campaigns of MIRI to simulate realistic point spread function (PSF), dark current, flat-fielding, detector read noise, Poisson noise, cosmic rays, etc. For the next six months following the JWST launch, instrument teams will compare real observations to MIRISim in order to adjust calibration data and check the performance of the instruments [9].

In this work, we use the illumin-images to which noise has been added, in order to obtain more realistic data instead of using the det-images which would imply using a pipeline. We perform simulations of a star-planet system, using spectra simulated with PySynphot and petitRADTRANS, respectively. In Section 5, we will explain the related procedure and the simulation of the objects of interest. We will assume that we do not have enough spacial resolution to distinguish the parent star from the planet, and so both objects will be considered as a point object located in the center of our observation scene.

4.2. Star Spectra Modeling: PySynphot

PySynphot is a package that simulates photometric data and spectra, allowing the generation of complicated composite spectra from atmospheric models and stellar spectrophotometric atlases (see PySynphot Documentation [19]). We have decided to use Phoenix model since it fulfills our main requirements, i.e., it must be able to generate spectrum of star/brown dwarfs, with a cloud model, on the "spectra width band" in which LRS-Slitless works (5 to 12 μm).

4.3. Exoplanets Spectra Modeling: petitRADTRANS

The free code *petitRADTRANS* (pRT) was used to model the spectra [23]. It is a Python package for calculating transmission and emission spectra of exoplanets for clear and cloudy planets. This code allows 1-D spectra to be modeled for different types of planets: terrestrial, super-Earths, mini-Neptunes, and hot-Jupiters. To do this, a radiative transfer object must be created.

This is achieved by defining the parameters: radius, wavelength range, P-T profile, the atmospheric composition, including molecules that produce continuous opacities and/or scattering, the mean molecular weight (MMW)... By varying these parameters, we study how the transmission spectrum changes. Thus, with these input data, we carry out the recovery process (see Section 6.2) to determine whether or not it is possible to perceive these changes in the spectrum. Although at the moment, we are studying a very simple model: isothermal P-T profile, constant chemical equilibrium in each atmospheric layer, cloudless atmosphere, and low resolution data, we hope to carry out a similar analysis in the future with more realistic models.

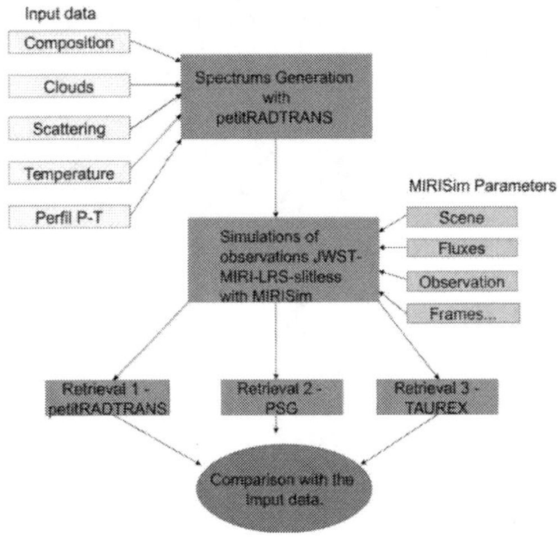

Figure 1: General workflow of the realized simulations.

5. System target and MIRISim observation scene

We use a star-planet system; a red dwarf like K2-18 and an hypothetical Super-Earth planet of 1.75 Earth radius (1.75 $R_{\oplus}$).

5.1. Star

In this work, the spectrum of a red dwarf was simulated using the following parameters: $T_{eff} = 3464K$, $\log g = 4.7864$ [27] and $[Fe/H] = 0.12$ (parameters were obtained from NASA Exoplanet Archive) [25].

5.2. Planet

We are interested in studying the variation of the transmission spectrum due to the change of the input parameters such as the equilibrium temperature and the concentration of the atmospheric gases, particularly the abundance of water vapor. Thus, after the retrieval process (see Section 6.2), the effects of these changes in the parameters should be noted in the spectra [14].

We studied two cases:

- Case-A: Variation of the water vapor concentration in the atmosphere dominated by H_2 and He, while keeping the equilibrium temperature ($400K$) constant, the planet's radius (1.75 $R_{\oplus}$), the planet's gravity on its surface ($Log(g) = 2.58$), and the pressure (see Table 1), with traces of other gases.

Table 1: Atmospheric compositions for Case-A. Concentrations are in volume mixing ratio (VMR).

Species	1% H_2O vapor	10% H_2O vapor	20% H_2O vapor
H_2	0.74	0.74	0.74
He	0.24	0.15	0.05
H_2O	0.01	0.10	0.20
CO_2	4×10^{-3}	4×10^{-3}	4×10^{-3}
CH_4	1×10^{-4}	1×10^{-4}	1×10^{-4}
CO	1×10^{-9}	1×10^{-9}	1×10^{-9}

In Figure 2 the three transmission spectra are shown, which are obtained varying the water vapor abundance.

- Case-B: The same parameters as in Case-A, except that the abundances of NH_4 and CO_2 have been increased (see table 2).

Table 2: Atmospheric compositions for Case-B. Concentrations are in volume mixing ratio (VMR).

Species	1% H_2O vapor	10% H_2O vapor	20% H_2O vapor
H_2	0.74	0.72	0.72
He	0.18	0.11	0.01
H_2O	0.01	0.10	0.20
CO_2	0.05	0.05	0.05
CH_4	0.016	0.016	0.016
CO	1×10^{-6}	1×10^{-6}	1×10^{-6}

The pRT retrieval works in mass fraction units for concentrations, but the retrieval returns the results as $log(X_{mol})$. The species we use are the most common species studied for exoplanets. On the one hand, for Case-A, we decided to use slightly higher abundances than what is generally used in the literature [14] to have sufficient abundance for its detection by pRT retrieval. On the other hand, for Case-B, we use an abundance of CH_4 similar to that observed on the Titan satellite [5]. The CH_4 increment in the atmosphere leads to an increment in CO for planets around M dwarfs [28]. And a higher CO concentration is necessary since CO_2 gas is produced directly from the recombination of CO and OH. In Figure 2 the three transmission spectra are shown.

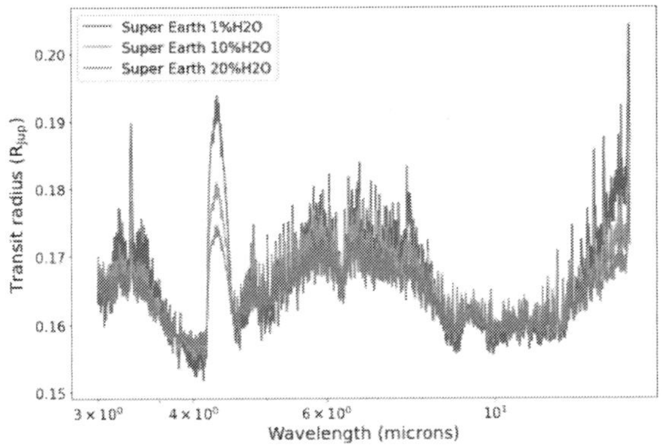

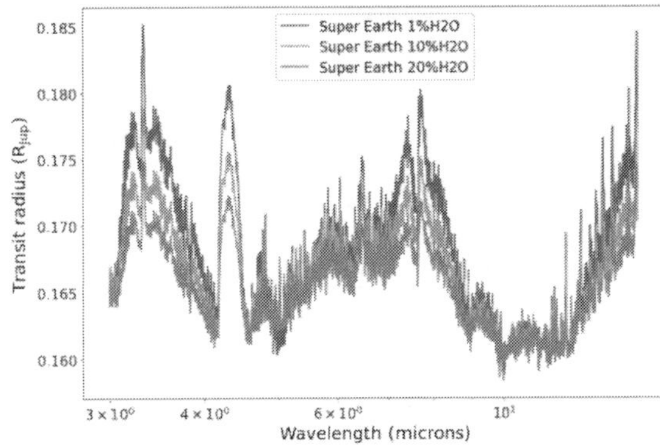

Figure 2: Comparison between transmission spectra by varying the abundance of water vapor in the atmospheric composition of a super-Earth-like planet (1.75 $R_{\oplus}$). The change of water vapor was balanced by the change in He. Left: Case-A; Right: Case-B.

5.3. Star-planet system

In order to simulate a star-planet system in such a way that it can be observed with the JWST, the simulations were performed with the following input parameters and spectra:

(i) A red dwarf with the same parameters described in Section 5.1.
(ii) The transit radius of a super-Earth planet generated in petitRADTRANS as specified in Sections 2 and 5.2.

Since petitRADTRANS gives the spectra in transit radius and MIRISim works with fluxes in μJy, to enter the flux of the planet-star system into MIRISim, the relation 1, indicated below was used, and thus, we obtained data ready to be treated in the subsequent stages of the simulation.

$$\frac{\Delta F}{F_S} = \left(\frac{R_P}{R_S}\right)^2, \tag{1}$$

where ΔF is the observed change in flux when the planet transits in front of the star. F_S and R_S are the flux and radius of the star, respectively. $R_P = R_P^{disk} + H$ is the wave-dependent radius of the planet which includes the contribution of stellar outflow blockage due to the atmosphere-less planetary disk (R_P^{disk}), and the atmospheric contribution described by the scale factor H. The transit depth corresponds to the left side of this equation, which is valid in the first order when the darkening of the limb is neglected. That is, it is assumed that the stellar disk has a uniform intensity [11]. Typically, super-Earths block stellar flow by about 1% or less, resulting in a transit depth to a maximum of 0.01% [7, 1].

5.4. MIRISim simulation parameters

We choose using the IMA mode with the configpath = $LRS - SLITLESS$. The IMA configuration had the following parameters: filter = P750L (double prism) that is the only one available for this mode; ReadDetect = SLITLESSPRISM and Mode = FAST, that maximizes the dynamic range and minimizes noise when the background is dominant. Finally, the number of exposures, frames and integrations are not relevant for the present study since the results that include detector effects will not be used.

6. Simulation results and data analysis

6.1. Analysis: Extraction of spectra from MIRISim fits images

The result of the simulations carries out with MIRISim are in a .fits file for each exposure, both for the illumination and detector (Figure 3a for illumination). The flux (in electrons/seconds) was extracted from the .fits file as a function of the wavelength (Figure 3b).

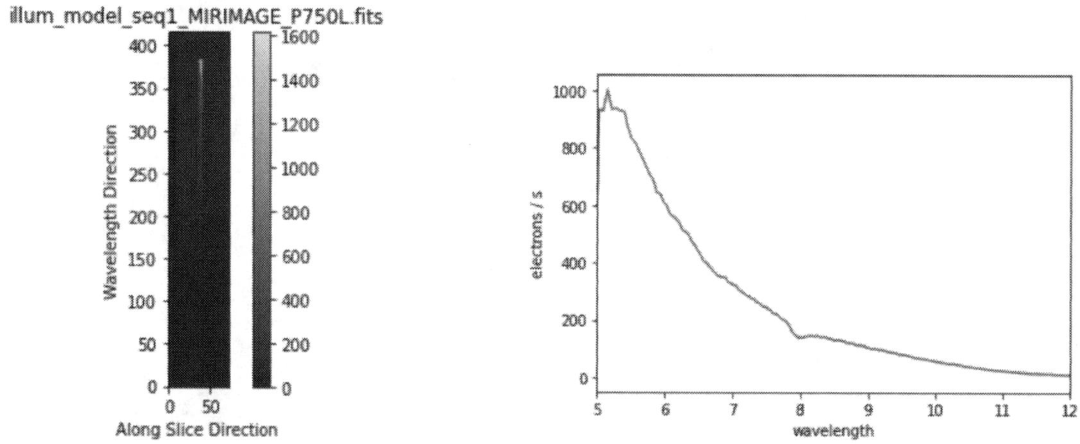

(a) (b)

Figure 3: Graphs of the illumination files obtained for the system: (a) Dispersion spectrum with respect to the pixels of the array for the illumination file. (b) Spectrum extracted from the FITS files.

Similar figures are obtained for the host star. From these results, the transit depth was calculated for Case-A, obtaining the curves of Figure 4, for both 1 % and 10 % concentration of H_2O, with the original spectra (before MIRISim). We plotted the same in Figure 5 but with the spectra obtained after MIRISim. One can see how the simulator reduces the resolution of the original spectra, since the resolution of MIRI-LRS-Slitless is very low, varying between $R = 60$ y $R = 150$ going from $5\mu m$ to $12\mu m$.

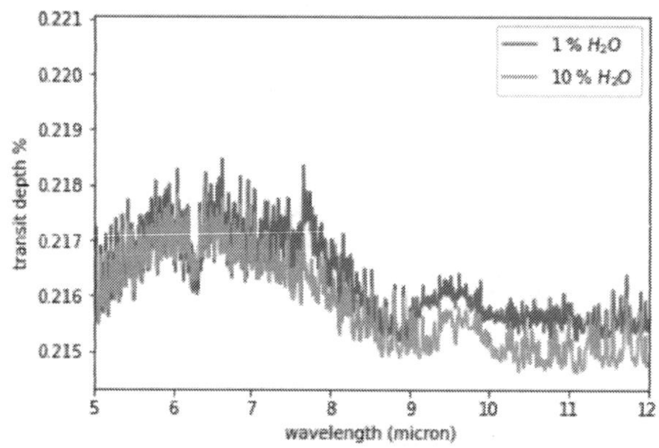

Figure 4: Transit depth obtained from the input data (before MIRISim) for both concentrations of water vapor in Case-A.

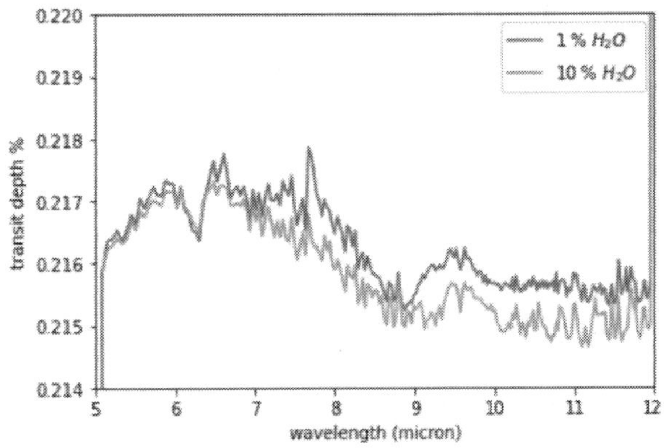

Figure 5: Transit depth retrieved from illumination results, relative to Case-A.

6.2. Retrieval with petitRADTRANS (pRT)

As we can see in Figure 1, different retrieval tools exist. For example, PSG, TauREX, and petitRADTRANS (pRT). In this work, we us pRT, but in the future, we pretend to compare the results of different codes.

The pRT retrieval process is based on a 1D atmospheric radiative transfer tool, combined with a sampler that allows us to statistically measure the properties of the studied planets. It uses nested sampling and Bayesian's tools to determine the parameter values that best fit the input data (see petitRADTRANS Retrieval Tutorial and [22], chapter 6).

We use as input the same spectra that were obtained from the spectra generation process with pRT. Data processing was done so that the units are the same as those required for the retrieval to work properly. Table 3 shows all the input parameters that we use to perform the various retrievals. There are fixed parameters like the radius of the star, which we assume to be a known value. Alternatively, the star radius can also be set as a free parameter, which was not the case in our study.

Table 3: Input parameters for the pRT retrieval process. The radius of the star is a fixed parameter.

Input Parameters	Values
Star Radius	0.469 $R_\odot$
Gravity Range (Log(g))	(1 to 5)
Planet Radius Range	(0.14 to 0.17) $R_\oplus$
Temperature Range	(150 to 1500) K

The molecules that were entered as input parameters were H_2O, CO_2, CH_4 and CO, while the molecules of H_2 and He were part of the parameters that cause Rayleigh scattering and continuous opacity of type $H_2 - H_2$ and $H_2 - He$. With the input parameters defined, we use the retrieval process integrated in pRT to find the best fit model (Figure 6) and the cornet plot (Figure 7) with the values that best fit our model.

The results of the retrieval process for the original spectra (non-modified by MIRISim) generated from pRT are shown in the figures: for 1% of water vapor in Figure 6 (best fit) and 10% of water vapor in Figure 7 (cornet plot), to assure retrieval tools are functional.

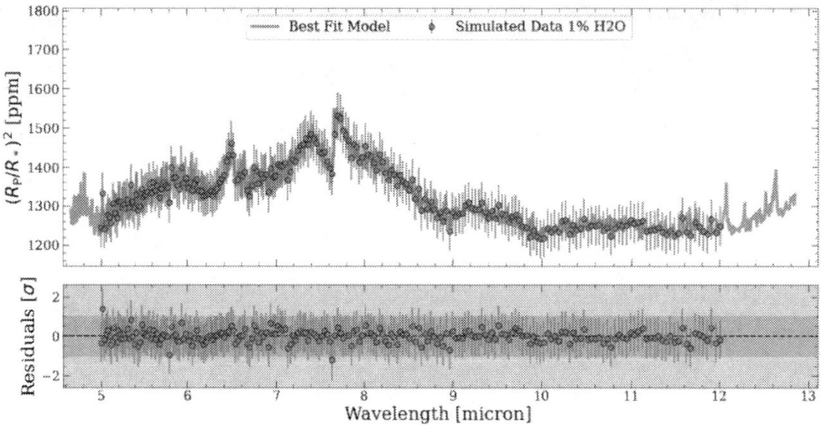

Figure 6: Best Fit Model for the case of the simulated planet with a 1% of H_2O. The input parameters are indicated in Table 3 and the atmospheric composition is in Table 1 (Case-A).

As we can see in Table 4, it is confirmed that the retrieval tool can detect the presence of different gases with a certain margin of error. In this table, we present the results of the spectral retrievals of the super-Earth with 20% H_2O , ranging from 3 to 15 μm and from 5 to 12 μm, compared to the parameters used to generate the synthetic spectrum. It is clear that retrieval over a wider range gives more accurate results, confirming that coupling NIRSPEC (0.6 to 5 μm) and MIRI-LRS-Slitless (5 to 12 μm) is important in exoplanets observations with JWST, but it is not the case for all the parameters. In theory, the MIRI-LRS-Slitless can operate from 4.5 μm and beyond 12 μm, but with a lot of noise. So, many studies focus only on the range of 5 to 12 μm.

Table 4: Table of the different retrieval results performed in different bands depending on the spectrum generation parameters (super-Earth, Case-A with 20% of H_2O).

Parameters	Retrieval 3-15 μm	Retrieval 5-12 μm	Model values
H_2O abundance	−0.443	−0.228	−0.445
CH_4 abundance	−3.660	−3.279	−3.497
CO_2 abundance	−1.771	−1.218	−1.455
CO abundance	−4.762	−3.830	−8.253
Surface gravity ($Log(g)$)	2.816	2.589	2.58
Planet radius (R_{Jup})	0.165	0.165	0.159
Temperature (K)	379.93	$342.56^{+24.8}_{-19.14}$	400

We observe that the values obtained are very close to those expected, especially those related to the radius of the planet and the temperature. Some discrepancies in the concentrations were observed. In the case of water at 20% the range found for H_2O abundance by the retrieval for 5-12 μm does not match the model value, and in the case of CO it does not match any value. These errors are most likely due to the range that the retrieval worked on to find these values. For the CO the concentrations were too small (on the order of 10^{-9}). However, the results for

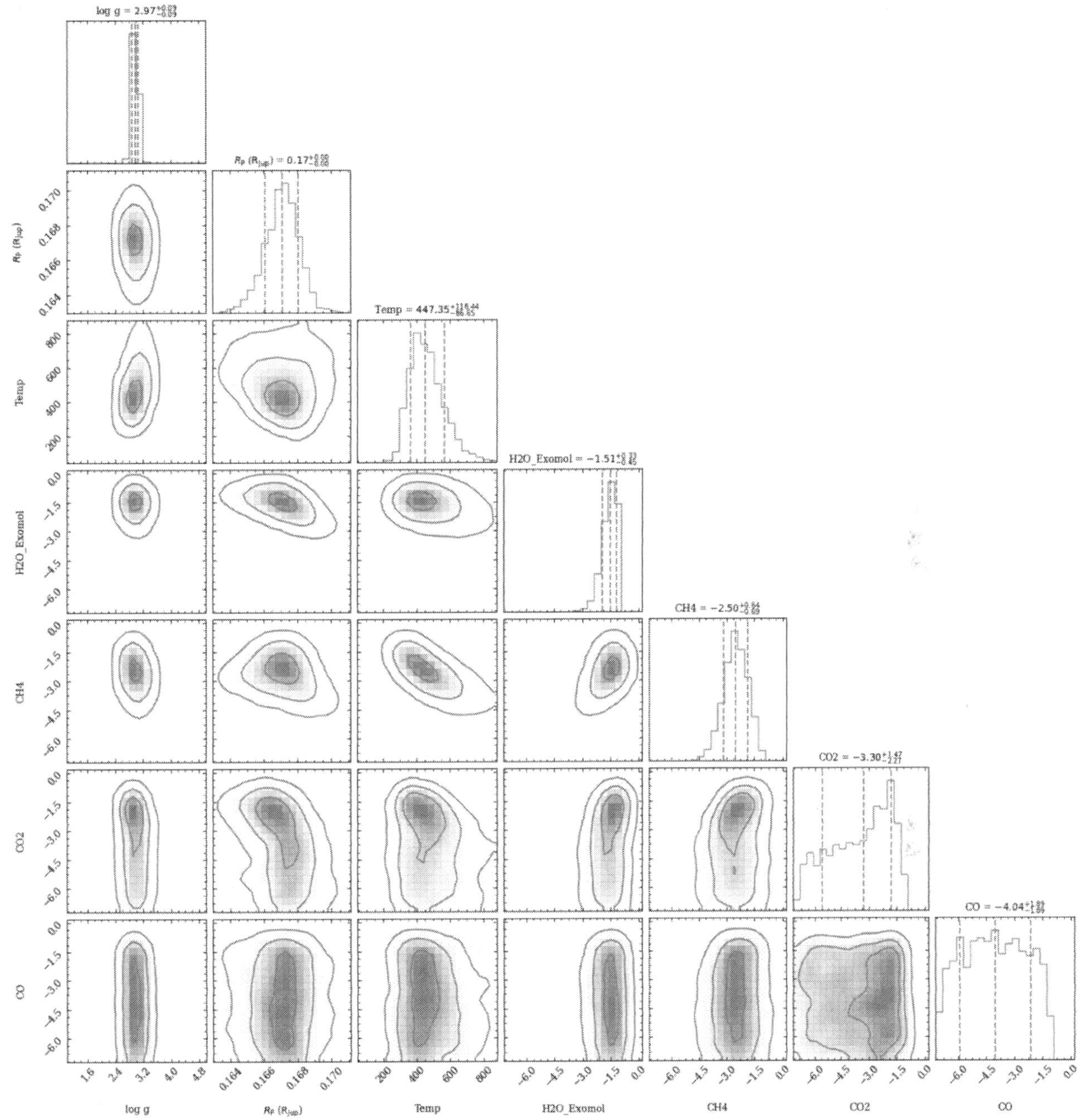

Figure 7: Corner graph of the planet retrieve process simulated with 10% H_2O. The most probable values for each parameter are shown at the top of each box. The abundance are in units of $log(X_{mol})$.

the abundances are more accurate when the elements have higher abundances as the cases for H_2O and CH_4.

6.3. Retrieval spectrum "Illumination"

The numerical results for the three compositions are in Table 5 and 6. The retrieval results of the data obtained from the MIRISim of the illumination part with added noise can be found in the results and discussion section.

Table 5: Cornet Plot parameter results compared to the model value: Case-A.

Parameters	Model Values	$1\% H_2O$	$10\% H_2O$	$20\% H_2O$
Surface gravity (Log(g))	2.58	$2.43^{+0.09}_{-0.07}$	$2.60^{+0.08}_{-0.08}$	$2.82^{+0.04}_{-0.05}$
Planet radius (R_{Jup})	0.156	$0.16^{+0.00}_{-0.00}$	$0.16^{+0.00}_{-0.00}$	$0.16^{+0.00}_{-0.00}$
Temperature (K)	400	$304^{+11.13}_{-9.68}$	$329.21^{+19.33}_{-14.47}$	$379.13^{+13.75}_{-14.41}$
H_2O abundance	$-2; -1; -0.699$	$-1.17^{+0.11}_{-0.10}$	$-0.39^{+0.09}_{-0.13}$	$-0.46^{+0.09}_{-0.09}$
CH_4 abundance	-4 *	$-2.96^{+0.07}_{-0.09}$	$-3.25^{+0.14}_{-0.15}$	$-3.68^{+0.05}_{-0.05}$
CO_2 abundance	-2.40 *	$-0.44^{+0.08}_{-0.16}$	$-1.10^{+0.16}_{-0.20}$	$-1.79^{+0.09}_{-0.10}$
CO abundance	-9 *	$-4.63^{+1.22}_{-0.90}$	$-4.04^{+1.37}_{-1.30}$	$-4.87^{+0.45}_{-0.66}$

* *This abundance is the same for the three different abundances of H_2O.*

Table 6: Cornet Plot parameter results compared to the model value: Case-B.

Parameters	Model Value	$1\% H_2O$	$10\% H_2O$	$20\% H_2O$
Surface gravity (Log(g))	2.58	$2.85^{+0.07}_{-0.07}$	$2.97^{+0.09}_{-0.09}$	$3.07^{+0.12}_{-0.10}$
Planet Radius (R_{Jup})	0.156	$0.17^{+0.00}_{-0.00}$	$0.17^{+0.00}_{-0.00}$	$0.17^{+0.00}_{-0.00}$
Temperature (K)	400	$419.40^{+76.82}_{-62.63}$	$447.35^{+118.44}_{-86.65}$	$481.05^{+162.48}_{-118.00}$
H_2O abundance	$-2; -1; -0.699$	$-2.00^{+0.42}_{-0.40}$	$-1.51^{+0.33}_{-0.45}$	$-1.52^{+0.34}_{-0.53}$
CH_4 abundance	-1.796 *	$-2.02^{+0.47}_{-0.54}$	$-2.50^{+0.64}_{-0.69}$	$-2.83^{+0.80}_{-0.84}$
CO_2 abundance	-1.301 *	$-1.66^{+0.48}_{-1.33}$	$-3.30^{+1.47}_{-2.27}$	$-4.08^{+1.68}_{-1.81}$
CO abundance	-6 *	$-3.96^{+1.90}_{-2.04}$	$-4.04^{+1.89}_{-1.89}$	$-4.12^{+1.89}_{-1.90}$

* *This abundance is the same for the three different abundances of H_2O.*

In the case of Surface gravity, a slight discrepancy with respect to the model value could be noted, especially in Case-B. However, the value of the radius of the planet was quite close in both cases. The reason a totally accurate result was not obtained is mainly because the atmosphere of the planets contributes to the size of the radius of the planet.

It was also observed that the temperature increased as the abundance of greenhouse gases, such as H_2O, CH_4 and CO_2, increased. The calculation of the abundance of water vapor in Case-A was not very successful, while in Case-B the abundance of water vapor was practically exact for the case of $1\% H_2O$. But in the other cases, the error was increasing. We assume that the relatively high presence of the other gases in the atmosphere makes it difficult to correctly determine each of the abundances. For this we conclude that it is necessary to compare the recovery process with different recovery tools such as TauREX or PSG.

6.4. Retrieval spectrum "Illumination plus noise"

In this section, we study the Case-B, e.i., with the composition of 1.6% of CH_4 and 5% of CO_2, with a 10% of water vapor, and adding noise. For the JWST, the Poisson noise is dominated by the background noise, so we added a Gaussian noise defined with different standard σ deviation (see JWST noise documentation).

Numerical results of the retrieval for Case-B, for the different compositions, are presented in Table 7 and best fits are shown in Figures 8a and 8b.

Table 7: Retrieval results for the super-Earth spectrum (Case-B) with Gaussian noise for three different values of σ.

Output Parameters	Model Value	$\sigma = 5 \times 10^{-5}$	$\sigma = 1 \times 10^{-4}$	$\sigma = 4 \times 10^{-4}$
Surface gravity ($Log(g)$)	2.58	$2.97^{+0.03}_{-0.02}$	$2.96^{+0.05}_{-0.05}$	$3.06^{+0.16}_{-0.17}$
Planet Radius (R_{Jup})	0.156	$0.16^{+0.00}_{-0.00}$	$0.16^{+0.00}_{-0.00}$	$0.16^{+0.00}_{-0.00}$
Temperature (K)	400	$433.23^{+25.44}_{-16.31}$	$454.4^{+44.91}_{-38.79}$	$442.85^{+199.67}_{-116.21}$
H_2O abundance	-1	$-1.04^{+0.03}_{-0.05}$	$-1.22^{+0.15}_{-0.25}$	$-2.3^{+0.65}_{-0.90}$
CH_4 abundance	-1.796 *	$-1.85^{+0.16}_{-0.15}$	$-2.18^{+0.28}_{-0.29}$	$-2.22^{+0.77}_{-1.06}$
CO_2 abundance	-1.301 *	$-1.26^{+0.15}_{-0.17}$	$-1.52^{+0.29}_{-0.40}$	$-4.37^{+2.03}_{-1.06}$
CO abundance	-6 *	$-4.33^{+1.91}_{-1.77}$	$-4.18^{+1.96}_{-1.95}$	$-3.89^{+1.77}_{-1.99}$

* *This abundance is the same for the three different levels of noise.*

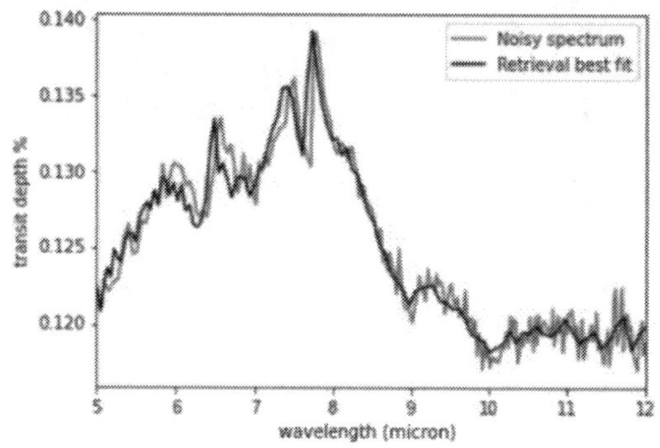

(a) Added noise with $\sigma = 5 \times 10^{-5}$.

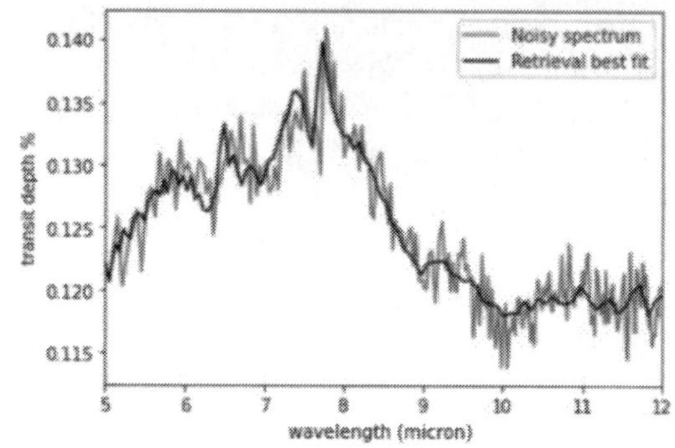

(b) Added noise with $\sigma = 4 \times 10^{-4}$.

Figure 8: Comparison between transmissions spectra with added noise and the best fit from the pRt retrieval.

As expected, as the noise increases, the retrieval results move further away from the model value. This can be seen by looking at Table 7. Therefore, if the noise exceeds a certain threshold ($\sigma = 0.0002$ in our simulations), the values obtained from the retrieval process will not present convincing results.

7. Discussion and conclusions

We have identified some general aspects to take into account in the use of MIRISim, such as: The default detector read mode is FASTR1, which contains a reset between integrations; The MIRISim still uses the FAST read mode, which does not include this additional reset. Therefore, the exposure time recorded in the simulated FITS header is missing this additional reset time.

MIRISim cannot insert a time-varying signal into a simulated observation. The flow of the source is assumed to be constant. Additional tools have been developed in the MIRI consortium for this purpose, but for the purposes of this work we do not consider it. Also, The MIRI LRS prism has a leak in its transmission causing a flux of 3-4 µm to contaminate the spectrum around 6-7 µm. This is not included in MIRISim models but it will be necessary take it into account in the future.

We also identified the following limiting factors in the spectroscopic characterization: the degradation of the retrieval results with the presence of noise shows the need to understand noise sources and their treatment through a pipeline. So, it would be useful in a future work to use the image of the detector that generates a more realistic noise than our simple Gaussian noise with MIRISim pipeline. The presence of more components such as CH_4 or CO_2 makes it more difficult to determine the composition of the atmosphere with the retrieval tool that we have used. In the future, we plan to compare different retrieval tools to see if the results can be improved or if we can combine them. The comparison retrieval with spectrum from 3 to 15 μm, and $5 - 12\mu$m shows that the results are better when the spectrum is larger. So, we recommend to make observation with MIRI and NIRSPEC to enhance the precision of the results.

Simulations have shown that MIRISim performs well for the study of super-Earth planets orbiting an M-class star. And petitRADTRANS, is seen as a useful radiative transfer package for simulating exoplanet spectra.

Bibliography

[1] E. D. Alonso, J. I. Hernández, B. Toledo-Padrón, S. L. Gómez, A. S. Mascareño, D. S. Aguado, C. G. Gutiérrez, A. Cabrera-Lavers, J. Carballido-Landeira, L. Bonavera, F. J. De Cos Juez, and R. Rebolo, *A transiting super-Earth close to the inner edge of the habitable zone of an M0 dwarf star*, Monthly Notices of the Royal Astronomical Society, 489 (2019), pp. 5928–5937.

[2] J. K. Barstow, S. Aigrain, P. G. Irwin, S. Kendre, and L. N. Fletcher, *Transit spectroscopy with James Webb Space Telescope: Systematics, starspots and stitching*, Monthly Notices of the Royal Astronomical Society, 448 (2015), pp. 2546–2561.

[3] C. A. Beichman and T. P. Greene, *Observing Exoplanets with the James Webb Space Telescope.*

[4] B. Benneke and S. Seager, *How to distinguish between cloudy mini-Neptunes and water/volatile-dominated super-Earths?*

[5] B. Benneke and S. Seager, *Atmospheric retrieval for super-Earths: uniquely constraining the atmospheric composition with transmission spectroscopy*, (2012).

[6] P. Bouchet, M. García-Marín, P.-O. Lagage, J. Amiaux, J.-L. Auguéres, E. Bauwens, J. A. D. L. Blommaert, C. H. Chen, Ö. H. Detre, D. Dicken, D. Dubreuil, P. H. Galdemard, R. Gastaud, A. Glasse, K. D. Gordon, F. Gougnaud, P. Guillard, K. Justtanont, O. Krause, D. Leboeuf, Y. Longval, L. Martin, E. Mazy, V. Moreau, G. Olofsson, T. P. Ray, J.-M. Rees, E. Renotte, M. E. Ressler, S. Ronayette, S. Salasca, S. Scheithauer, J. Sykes, M. P. Thelen, M. Wells, D. Wright, and G. S. Wright, *The Mid-Infrared Instrument for the James Webb Space Telescope, III: MIRIM, The MIRI Imager*, (2015).

[7] Cáceres, C., Kabath, P., Hoyer, S., Ivanov, V. D., Rojo, P., Girard, J. H., Miller-Ricci Kempton, E., Fortney, J. J., and Minniti, D., *Ground-based transit observations of the super-earth gj4*, A&A, 565 (2014), p. A7.

[8] E. Chiang and G. Laughlin, *The minimum-mass extrasolar nebula: in situ formation of close-in super-Earths*, MNRAS, 431 (2013), pp. 3444–3455.

[9] M. E. Consortium, *MIRISim Documentation*, (2018).

[10] C. Danielski, J.-L. Baudino, P.-O. Lagage, A. Boccaletti, R. Gastaud, A. Coulais, and B. Bézard, *Atmospheric Characterization of Directly Imaged Exoplanets with JWST /MIRI*, The Astronomical Journal, 156 (2018), p. 276.

[11] D. Deming, D. Louie, and H. Sheets, *How to characterize the atmosphere of a transiting exoplanet*, Publications of the Astronomical Society of the Pacific, 131 (2019), p. 013001.

[12] J. P. Gardner, J. C. Mather, M. Clampin, R. Doyon, M. A. Greenhouse, H. B. Hammel, J. B. Hutchings, P. Jakobsen, S. J. Lilly, K. S. Long, J. I. Lunine, M. J. Mccaughrean, M. Mountain, J. Nella, G. H. Rieke, M. J. Rieke, H.-W. Rix, E. P. Smith, G. Sonneborn, M. Stiavelli, H. S. Stockman, R. A. Windhorst, and G. S. Wright, *The james webb space telescope*, Space Science Reviews, 123 (2006), pp. 485–606.

[13] A. Glasse, G. H. Rieke, E. Bauwens, M. García-Marín, M. E. Ressler, S. Rost, T. V. Tikkanen, B. Vandenbussche, and G. S. Wright, *The Mid-Infrared Instrument for the James Webb Space Telescope , IX: Predicted Sensitivity*, Publications of the Astronomical Society of the Pacific, 127 (2015), pp. 686–695.

[14] T. P. Greene, M. R. Line, C. Montero, J. J. Fortney, J. Lustig-Yaeger, and K. Luther, *Characterizing Transiting Exoplanet Atmospheres With Jwst*, The Astrophysical Journal, 817 (2016), p. 17.

[15] M. Greenhouse, *The jwst science instrument payload: Mission context and status*, vol. 9143, 06 2014.

[16] E. P. Jan, *Simulated jwst/niriss transit spectroscopy of anticipated tess planets compared to select discoveries from space-based and ground-based surveys*, (2018).

[17] C. M. J.M. Trigo-Rodriguez, François Raulin, *Evolution of the atmospheres of the terrestrial planets*, 2013.

[18] P. D. Klaassen, V. C. Geers, S. M. Beard, A. D. O'brien, C. Cossou, R. Gastaud, A. Coulais, J. Schreiber, P. J. Kavanagh, M. Topinka, R. Azzollini, W. De Meester, J. Bouwman, A. C. H. Glasse, A. M. Glauser, D. R. Law, M. Cracraft, K. Murray, B. Sargent, O. C. Jones, and G. S. Wright, *MIRISim: A Simulator for the Mid-Infrared Instrument on JWST*, tech. rep., 2020.

[19] P. L. Lim, R. I. Diaz, and V. Laidler, *Pysynphot user's guide*, (2015).

[20] E. D. Lopez and J. J. Fortney, *Understanding the mass-radius relation for sub-Neptunes: radius as a proxy for composition*, (2013).

[21] R. G. Martin and M. Livio, *On the formation of super-Earths with implications for the solar system*, The Astrophysical Journal, 822 (2016).

[22] P. Mollière, *Modeling of exoplanet atmospheres*, (2017).

[23] P. Mollière, *petitradtrans documentation*, (2019).

[24] P. Mollière, J. P. Wardenier, R. van Boekel, T. Henning, K. Molaverdikhani, and I. A. G.

Snellen, *petitRADTRANS: a Python radiative transfer package for exoplanet characterization and retrieval*, 67 (2019), pp. 1–17.

[25] P. Sarkis, T. Henning, M. Kürster, T. Trifonov, M. Zechmeister, L. Tal-Or, G. Anglada-Escudé, A. P. Hatzes, M. Lafarga, S. Dreizler, I. Ribas, J. A. Caballero, A. Reiners, M. Mallonn, J. C. Morales, A. Kaminski, J. Aceituno, P. J. Amado, V. J. S. Béjar, H.-J. Hagen, S. Jeffers, A. Quirrenbach, R. Launhardt, C. Marvin, and D. Montes, *The CARMENES Search for Exoplanets around M Dwarfs: A Low-mass Planet in the Temperate Zone of the Nearby K2-18*, , 155 (2018), p. 257.

[26] E. Schlawin, T. P. Greene, M. Line, J. J. Fortney, and M. Rieke, *Clear and Cloudy Exoplanet Forecasts for JWST: Maps, Retrieved Composition and Constraints on Formation with MIRI and NIRCam*, (2018).

[27] K. G. Stassun, R. J. Oelkers, M. Paegert, G. Torres, J. Pepper, N. De Lee, K. Collins, D. W. Latham, P. S. Muirhead, J. Chittidi, B. Rojas-Ayala, S. W. Fleming, M. E. Rose, P. Tenenbaum, E. B. Ting, S. R. Kane, T. Barclay, J. L. Bean, C. E. Brassuer, D. Charbonneau, J. Ge, J. J. Lissauer, A. W. Mann, B. McLean, S. Mullally, N. Narita, P. Plavchan, G. R. Ricker, D. Sasselov, S. Seager, S. Sharma, B. Shiao, A. Sozzetti, D. Stello, R. Vanderspek, G. Wallace, and J. N. Winn, *The Revised TESS Input Catalog and Candidate Target List*, , 158 (2019), p. 138.

[28] F. Wunderlich, M. Godolt, J. L. Grenfell, S. Städt, A. M. Smith, S. Gebauer, F. Schreier, P. Hedelt, and H. Rauer, *Detectability of atmospheric features of Earth-like planets in the habitable zone around M dwarfs*, Astronomy and Astrophysics, 624 (2019), pp. 1–18.

Ionospheric total electron content (TEC) above Ecuador

E D Lopez Izurieta[1,2], **E Toapanta Guamanarca** [1], **H Barbier**[1]

[1] Escuela Politécnica Nacional, Facultad de Ciencias, Departamento de Física, Quito, Ecuador
[2] Escuela Politécnica Nacional, Observatorio Astronómico de Quito, Quito, Ecuador

E-mail: ericsson.lopez@epn.edu.ec

Abstract. The equatorial ionosphere over the Ecuadorian territory has not been previously studied, although there are some studies in other equatorial geographical positions, local research is required to characterize the ionospheric magnetoactive plasma in this region.

In this study, for the first time in Ecuador, using pseudophase and pseudorange data from the satellite signals of the GPS system, detected at the Riobamba-Ecuador (RIOP) station, the Total Electronic Content (TEC) is estimated in the period between the years 1999 and 2019. The TEC values and their daily, weekly, monthly and annual variation are calculated and compared with the observed data of the IGS network (International GNSS Service). The fit of the observed series with the series of the dispersive model has a goodness of fit of 0.88, which is quite acceptable. The adjustment can be improved by incorporating data from new local stations, as well as instrumental errors and delays and in the received signals, all of which were not considered in the present work.

1. Introduction

The total electron content (TEC) is a parameter that quantifies the Earth's ionosphere state, that influences on the propagation of electromagnetic waves which pass through it [15, 21]. Measuring the total electron content of the ionosphere is important in the investigation of ionospheric morphology, since innumerable physical processes take place in the upper atmosphere. One tool used to measure TEC is the global positioning system (GPS). GPS is a satellite-based radio navigation system that is used to verify the position in real time with an error of a few centimeters [21]. An important aspect of TEC's study is that allows a better understanding of the spatial and temporal variation of the ionosphere, thus allowing the implementation of new tools to make more precise ionospheric corrections in satellite communication and navigation systems.

The Earth's ionosphere is a layer of the atmosphere that ranges from 100 to 1000 km in height [10]. Its name is due to the fact that the particles that make it up are ionized, that is, positively and negatively charged. The ionization of the particles (atoms, molecules, ions and electrons) present in the ionosphere occurs because the particles are directly exposed to solar radiation [9, 4].

The equatorial zone of the ionosphere has the highest values of electron density. Solar radiation, electric and magnetic fields increase the number of electrons and the movement of them along the horizontal lines of the magnetic field. This movement causes a displacement of electrons to geomagnetic latitudes between 10 to 20 degrees, causing a high concentration of electrons in that zone (equatorial electro-jets) [12].

In this work we study for the first time the ionosphere over the Ecuadorian region, characterizing it with the study of the total content of electrons (TEC), using data from local stations, in particular from the Riobamba station, which has the most complete and reliable data. These types of studies have not been carried out before in Ecuador and are necessary for the correction of delays and degradation of the radio signals that we receive from the GPS global positioning system. The study of the TEC and its variability is of vital importance and very useful to minimize errors when high precision is required in the measurements of the global navigation satellite system (GNSS). This article presents the methodology used for the calculation of the total electron content of the equatorial ionosphere for a period from 1999 to 2019, using the carrier phase and pseudorange parameters, which were obtained from a NASA archive for the Riobamba-Ecuador station of the GPS satellite system. Dual frequency observations can be used to measure the ionosphere delay. This delay can then be removed from the measurements by combining the satellite frequencies $L_1 = 154\ f_0$ and $L_2 = 120\ f_0$, being $f_0 = 10.23$ MHz, the fundamental frequency of GPS system. We present in this contribution these novel results of the ionospheric TEC about the Ecuador.

2. Ionosphere: dispersive medium properties

The ionosphere acts as a refractive medium to radio signals, and the index of refraction depends on the amount of ionisation. The ionosphere affects a wide range of frequencies, from extremely low frequencies (about 3 kHz) to super high frequencies (30 GHz) [3].

The behavior of electromagnetic waves in the ionosphere can be described by considering that the ionosphere is a plasma subject to a non-uniform magnetic field and disturbances caused by solar wind [8]. The ionosphere is then considered as a magneto-active medium whose refractive index is a complex number [6]. The complex refractive index n is given by the Appleton-Hartree equation (see e.g., Komjathy [1997]):

$$n^2 = 1 - \frac{X}{(1-jZ) - \left[\frac{Y_T^2}{2(1-X-jZ)}\right] \pm \left[\frac{Y_T^4}{4(1-X-jZ)^2} + Y_L^2\right]^{1/2}} \tag{1}$$

where:

$$X = \frac{\omega_N^2}{\omega^2} = \frac{f_N^2}{f^2} \tag{2}$$

$$Y = \frac{\omega_H}{\omega} = \frac{f_N}{f} \tag{3}$$

$$Y_L = \frac{\omega_L}{\omega}, \quad Y_T = \frac{\omega_T}{\omega}, \quad and \quad Z = \frac{\omega_C}{\omega} \tag{4}$$

The angular frequencies are defined as:

ω is the angular frequency of the exploring wave.

ω_C is the angular collision frequency between electrons and heavier particles.

ω_N is the angular plasma frequency, $\omega_N^2 = \dfrac{Ne^2}{\epsilon_o m}$, with electron density N, electron charge e, permittivity of free space ϵ_o and electron mass m.

ω_H is the cyclotron angular frequency , $\omega_H = \dfrac{B_o|e|}{m}$, with the intensity of magnetic field B_o.

ω_L is the longitudinal component of the cyclotron angular frequency, with $\omega_L = \dfrac{B_o|e|}{m} cos\Theta$.

ω_T is the transverse component of the cyclotron angular frequency, with $\omega_T = \dfrac{B_o|e|}{m} sin\Theta$.

For the case, when the collisions are negligible, we have $Z = 0$ and the refraction index becomes:

$$n^2 = 1 - \frac{2X(1-X)}{2(1-X) - Y_T^2 \pm \left[Y_T^4 + 4(1-X)^2 Y_L^2\right]^{1/2}} \tag{5}$$

According to the magneto-optic theory, a plane polarized electromagnetic wave will split into two wave components: an ordinary wave (sign "+" in the previous equation) and an extraordinary wave (sign "−").

Equation (5) can be expanded by a Taylor series in terms of $\frac{1}{f}$ as:

$$n \cong 1 - \frac{1}{2}X \pm \frac{1}{2}XY|cos\Theta| - \frac{1}{8}X^2 - \frac{1}{4}XY^2(1 + cos^2\Theta) + ... \tag{6}$$

Equation (6) is valid when the frequency of the electromagnetic wave is higher than the plasma, collision and cyclotron frequencies [2]. This approach is known as quasi-longitudinal.

When both collisions and the magnetic field are negligible, we only use the first and second terms of Eq. (6) and the refraction index for the ionospheric plasma, that we adopt to calculate the TEC, is simply expressed as:

$$n = 1 - \frac{1}{2}X = 1 - \frac{Ne^2}{2\epsilon_o m\omega^2}. \tag{7}$$

2.1. Ionospheric modelling using GPS data

For the ionospheric modeling, refraction in the ionosphere is considered as one of the main sources of error in GPS positioning [3]. This refraction causes delays in signal propagation. The total electron content (TEC) between point A and point B is defined as the total number of free electrons contained in a column (1 m^2 cross section) joining A and B. The TEC is measured in TECU units (1 $TECU = 10^{16} \; \frac{e}{m^2}$) and for its estimation, the carrier phase and pseudorange data are used, which are provided by the GPS-TEC scintillation monitors. Therefore, the TEC is calculated by:

$$TEC = \int_A^B N(s)ds, \tag{8}$$

where N is the electron density (in $\frac{elect}{m^3}$), ds is the path integral differential and the integral is calculated along the path that joins points A and B. In the case of being a vertical column we will have TEC vertical (TECv), that is, in the direction of the zenith.

The delay that a signal experiences when it passes through the ionosphere, it depends on the density of electrons presents in the medium [19]. The ionospheric delay is defined as follows:

$$d_{ion} = \int_{Sat}^{Rec} (n-1)ds, \tag{9}$$

where Rec is the reception point and Sat the satellite one.

The equation (7) is used to replace the refraction index into the equation (9) and find the ionospheric delay as a function of the TEC:

$$\begin{aligned} d_{ion} &= \int_{Sat}^{Rec} \left(1 - \frac{Ne^2}{2\epsilon_o m\omega^2} - 1\right) ds \\ d_{ion} &= \frac{e^2}{2\epsilon_o m\omega^2} TEC. \end{aligned} \tag{10}$$

In the development of the previous equation was also used the equation (8) to eliminate the path integral and express the ionospheric delay as a function of TEC and frequency. Finally, using the typical values for the constants such as the electron charge ($e = 1.6 * 10^{-19}$ coulomb), permittivity of free space ($\epsilon_o = 8.8542 * 10^{-12}$ farad/m), electron mass ($m = 9.1095 * 10^{-31}$ kg) and with the fact that the angular frequency is $\omega = 2\pi f$, the simplified expression for ionospheric delay is found to be:

$$d_{ion} = \frac{40.3}{f^2} \; TEC \tag{11}$$

2.2. Ionospheric Observable

The GPS pseudorange (P) and carrier phase (Φ), observable parameters, are given, for the two GPS frequencies, by [11, 20, 1] :

$$P_1 = \rho + c(dT - dt) + d_{ion,L_1} + d_{trop} + \eta_{P_1} + \epsilon_{P_1} \tag{12}$$

$$P_2 = \rho + c(dT - dt) + d_{ion,L_2} + d_{trop} + \eta_{P_2} + \epsilon_{P_2} \tag{13}$$

$$\Phi_1\lambda_1 = \rho + c(dT - dt) + \lambda_1 N_1 - d_{ion,L_1} + d_{trop} + \eta_{\Phi_1} + \epsilon_{\Phi_1} \tag{14}$$

$$\Phi_2\lambda_2 = \rho + c(dT - dt) + \lambda_2 N_2 - d_{ion,L_2} + d_{trop} + \eta_{\Phi_2} + \epsilon_{\Phi_2} \tag{15}$$

where:
P_1, P_2 are the pseudorange measurements on L_1 and L_2, respectively.
Φ_1 , Φ_2 are the carrier phase observations on L_1 and L_2, respectively.
λ_1 , λ_2 are the wavelengths of the L_1 and L_2 carriers, respectively.
N_1 , N_2 are the carrier unknown phase ambiguities[1] on L_1 and L_2 , respectively.
η_{P_1}, η_{P_2} are the instrumental delays on P_1 and P_2, respectively.
ϵ_{P_1}, ϵ_{P_2} are errors produced by the noise of the received signals on P_1 and P_2, respectively.
η_{Φ_1}, η_{Φ_2} are the instrumental delays on Φ_1 and Φ_2, respectively.
ϵ_{Φ_1}, ϵ_{Φ_2} are errors produced by the noise of the received signals on Φ_1 and Φ_2, respectively.
d_{ion,L_1}, d_{ion,L_2} are the ionospheric delays corresponding to the frequencies L_1 y L_2, respectively.
ρ is the geometric range between the satellite and the receiver,
c is the vacuum speed of light,
dT, dt are the offsets of the receiver and satellite clocks from GPS time, respectively.
d_{trop} is tropospheric delay.

After subtracting the equation (12) of (13) and the equation (14) of (15), rearranging the equations, we get:

$$P_1 - P_2 = d_{ion,L_1} - d_{ion,L_2} + (\eta_{P_1} - \eta_{P_2} + \epsilon_{P_1} - \epsilon_{P_1}) \tag{16}$$

$$\Phi_1\lambda_1 - \Phi_2\lambda_2 = -d_{ion,L_1} + d_{ion,L_2} + (\eta_{\Phi_1} - \eta_{\Phi_2} + \epsilon_{\Phi_1} - \epsilon_{\Phi_2}) + (\lambda_1 N_1 - \lambda_2 N_2) \tag{17}$$

The contribution of terms in parentheses for the precision of TEC estimation is minimal, so they can be neglected ($\eta_{P_1} - \eta_{P_2} + \epsilon_{P_1} - \epsilon_{P_1} \approx 0$ and $\eta_{\Phi_1} - \eta_{\Phi_2} + \epsilon_{\Phi_1} - \epsilon_{\Phi_2} + \lambda_1 N_1 - \lambda_2 N_2 \approx 0$).

Since ionospheric delay cannot be measured directly, it has to be estimated. For this purpose, equation (11) is used, which expresses the ionospheric delay in terms of TEC and the frequency of the carrier wave. After combining equations (16) and (11), the expression for the total electron content using pseudorange observations is obtained as follows:

$$\begin{aligned} P_1 - P_2 &= d_{ion,L_1} - d_{ion,L_2} \\ P_1 - P_2 &= 40.3\ TEC_P \left(\frac{f_2^2 - f_1^2}{f_1^2 f_2^2}\right) \end{aligned} \tag{18}$$

$$TEC_P = \frac{(P_1 - P_2)}{40.3}\left(\frac{f_1^2 f_2^2}{f_2^2 - f_1^2}\right). \tag{19}$$

Similarly, after combining equations (17) and (11), the expression for the total electron content using carrier phase observations is obtained as follows:

$$\begin{aligned} \Phi_1\lambda_1 - \Phi_2\lambda_2 &= d_{ion,L_1} - d_{ion,L_2} \\ \Phi_1\lambda_1 - \Phi_2\lambda_2 &= -40.3\ TEC_\Phi \left(\frac{f_2^2 - f_1^2}{f_1^2 f_2^2}\right) \end{aligned} \tag{20}$$

$$TEC_\Phi = -\frac{(\Phi_1\lambda_1 - \Phi_2\lambda_2)}{40.3}\left(\frac{f_1^2 f_2^2}{f_2^2 - f_1^2}\right). \tag{21}$$

Replacing the values of GPS satellite frequencies ($f_1 = 1575.42\ MHz$) and ($f_2 = 1227.60\ MHz$) in equations (19) and (21), we get the simplified expressions of the TEC.

$$TEC_P = 9.52 \cdot (P_2 - P_1) \tag{22}$$

$$TEC_\Phi = 9.52 \cdot (\Phi_1\lambda_1 - \Phi_2\lambda_2). \tag{23}$$

In order to take full advantage of the information obtained from the ionospheric observable parameters; the pseudorange and carrier phase, the two expressions of the TEC are combined into a single expression [14]:

$$TEC_{comb_i} = TEC_{\Phi_i} - \frac{\sum_{j=1}^{n} p_j \left[TEC_{\Phi_j} - TEC_{P_j}\right]}{\sum_{j=1}^{n} p_j}, \tag{24}$$

where i and j are indices for ionospheric observables (i, j = 1,2,..., n), n is the total number of observations. The variable $p_j = \frac{1}{j^2}$ acts as a weighting factor, establishing the individual weight of each TEC measurement (TEC_{P_i} or TEC_{Φ_i}). Finally, the vertical TEC is calculated, which is a measure of the total electron content in a column perpendicular to the ground.

$$TEC_v = TEC_{comb}\,(cos\chi'), \tag{25}$$

where χ' is the difference between 90 ^{o} and the zenith angle [7] ($\chi' = 90^o - z$). The value of χ' is obtained from the relation:

$$sin\chi' = \frac{R_e}{R_e + h_m} sin(z), \tag{26}$$

where R_e is the earth radius and h_m is the height in which the highest electron density is found. Typical values of R_e and h_m are 6371 km and 506.7 km, respectively[17, 18].

According to the equation (24), the TEC can be calculated by unifying the data of pseudorange (P_1 and P_2) and carrier phase ($\Phi_1\lambda_1$ and $\Phi_2\lambda_2$). However, at the time of collecting the data, many problems related to the carrier phase arose, specifically irregularities were found with respect to the constancy in which the data were presented. Being more specific, there were time intervals of days, weeks and even months in which no carrier phase data was recorded. Thus, it was decided to use only the pseudorange data to calculate the TEC. In this way, the equation (24) becomes simply an average over the TEC_P, since the factor p_j becomes a constant and the TEC_Φ is not considered.

$$\begin{aligned} TEC_{comb_i} &= TEC_{\Phi_i} - \frac{\sum_{j=1}^{n} p_j \left[TEC_{\Phi_j} - TEC_{P_j}\right]}{\sum_{j=1}^{n} p_j} \\ TEC_{comb_i} &= \frac{\sum_{j=1}^{n} \left[TEC_{P_j}\right]}{n}. \end{aligned} \tag{27}$$

3. Estimation of the equatorial TEC

The characterization of the ionosphere in Ecuador has not been previously carried out in this region. In this sense, it is a very important pioneering project that allows for the first time to characterize the ionosphere by estimating the total content of electrons, corresponding to this geographical latitude. For this work, since the ionospheric TEC is going to be calculated in the equatorial zone, the daily files of the GPS receivers located in Ecuador were used. Specifically, the receiver located in Riobamba called RIOP, whose latitude and longitude are -1.65 and -78.65, respectively. This receiver was chosen because it is the oldest (beginning its operation in 1998),

 2238 (2022) 012010 doi:10.1088/1742-6596/2238/1/012010

and a long time series can be analyzed. The data extracted from these files, for the calculation of the TEC, are those of pseudorange(P_1, P_2) and phase (L_1, L_2)[13].

The data files are located in a directory of the IGS network website (International GNSS Service). This page is attached to the official page of NASA, of where you can access the daily records of the measurements made by the satellites and receivers all over the world. The data is given in the RINEX (Receiver INdependent EXchange) standard format for receivers of satellite navigation systems (GPS, GLONAS, Galileo).

Then, using the relations described above, the TEC for the equatorial ionosphere over the Riobamba station, has been calculated. The Figure 1 shows the daily, weekly and annual behavior of the total electron content for the equatorial ionosphere:

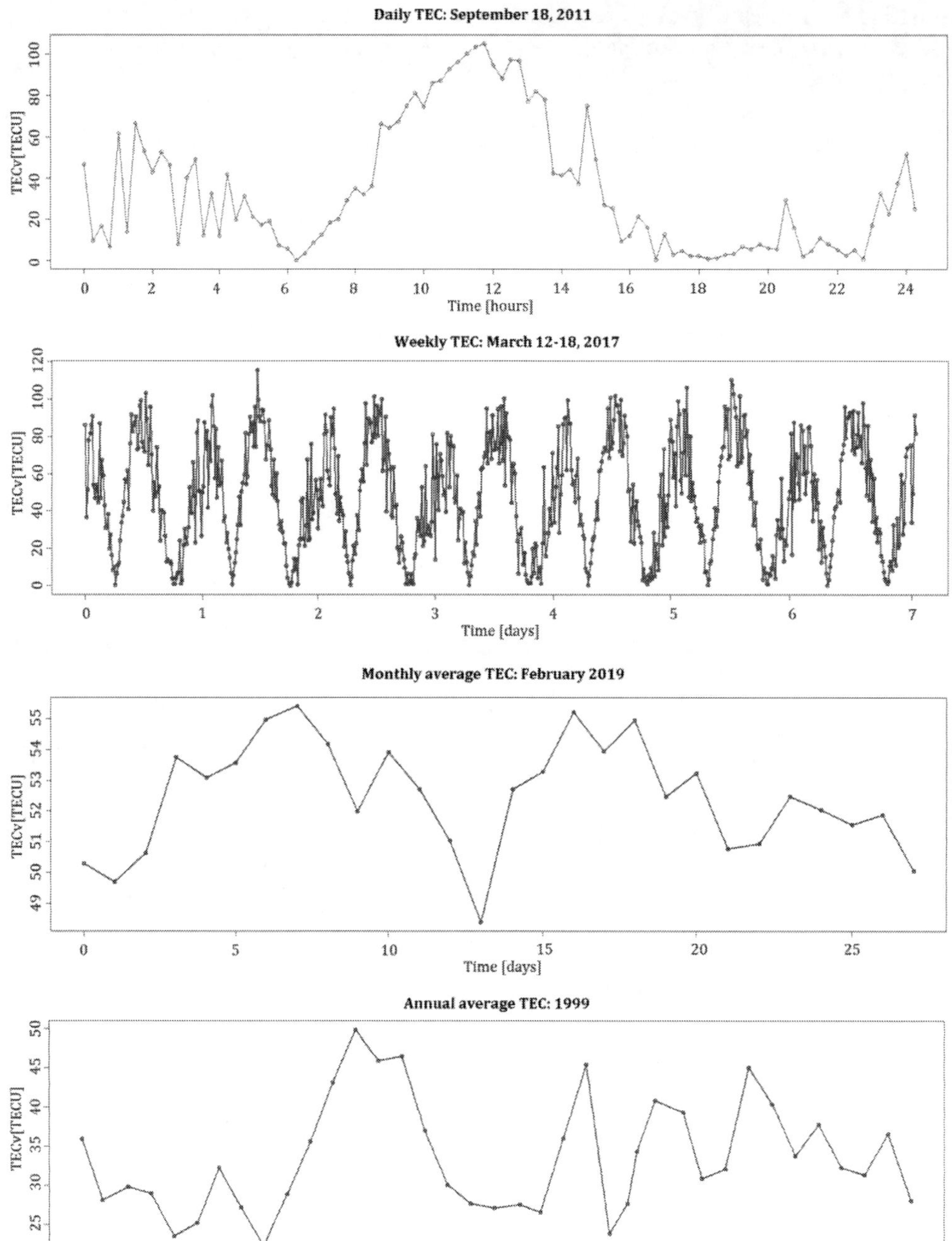

Figure 1: The figure shows the daily, weekly, monthly and annual TEC of randomly chosen dates. Pseudorange and carrier phase data were used, as well as the equations 22, 23, 24 and 25 to calculate the TECv.

In the Figure 2 the cumulative average TEC for the whole period from 1999 to 2019 is presented.

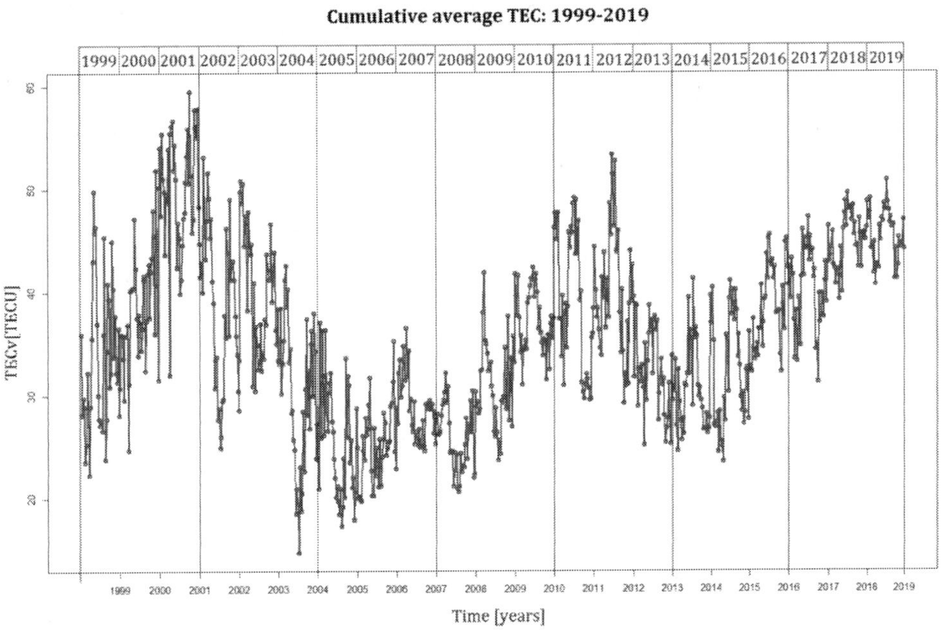

Figure 2: The figure shows the cumulative average TEC.

4. Results

Figure 1 shows results of TEC at different time intervals. For the daily TEC there is an increase in the total electron content around noon. As for the weekly TEC, the behavior is as expected since it is only the accumulated TEC of several days, it can be seen that there is an increase or decrease in the TEC depending on the time of day. In the monthly TEC there is no specific behavior, it can be seen that there is an increase or decrease in the TEC on certain days of the month. For the annual TEC it can be noted that there is an increase in the total electron content between the months of March and April, then in the months of September and October. This is directly related to the two annual equinoxes since at that time the Sun is at its closest point to the equator. It can also be noted that there is a decrease in the total electron content between the months of June and July, as well as in the months of January and December. Similarly, this is directly related to the two annual solstices since on these dates the Sun is at its farthest point from the equator.

The most general result was shown in Figure 2, it can be seen that there is an increase or decrease in different periods. It is important to mention that the TEC is affected by solar activity (which depends on the solar cycle) on the surface of the Sun, in this way it can be seen that the increase in the total content of electrons occurs precisely in the years close to solar maximums (2001, 2002, 2013 and 2014). On the other hand, the TEC decreases in the years close to the solar minimums (2008 and 2009).

5. TEC calculated vs. observed

The TEC values estimated with the dispersion relations described above (calculated values), can be compared with the values given by the IGS network (observed values). The Figure 3 show the comparison of the observed TEC (www.igs.org) and the calculated ones, with the data from the Riobamba station, for the period from 1999 to 2019.

They both have the same behavior and are generally very similar. However, after a more rigorous analysis it was found that there is an average relative error of 16.08 %. This means that the model used for estimating the total electron content in the equatorial ionosphere works quite well. The chi square test ($\chi^2 = \sum \frac{(O_i - C_i)^2}{C_i}$; O_i observed values, C_i calculated values) of the fit of the data with the dispersive model gives a goodness of the fit about 0.88, so the fitting could be considerate acceptable. The fit can be improved by incorporating data from new local stations, as well as instrumental errors and delays and in the received signals, the same which were not considered in this work.

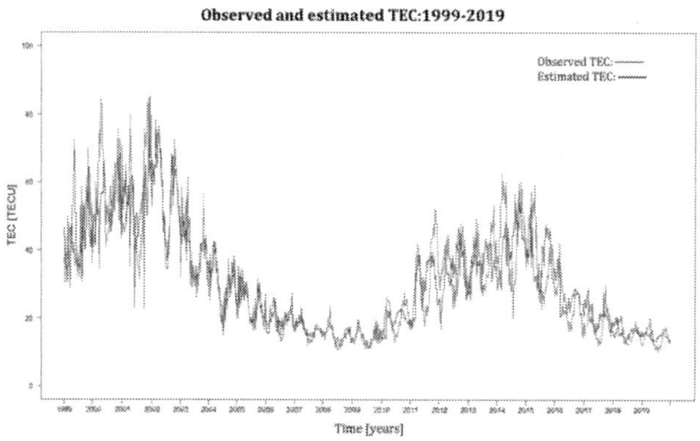

Figure 3: The observed and estimated TEC from 1999 to 2019.

In the Figure 4, the TEC for Ecuador with the respective error bar is given, for the ten-years period from 1999 to 2019 .

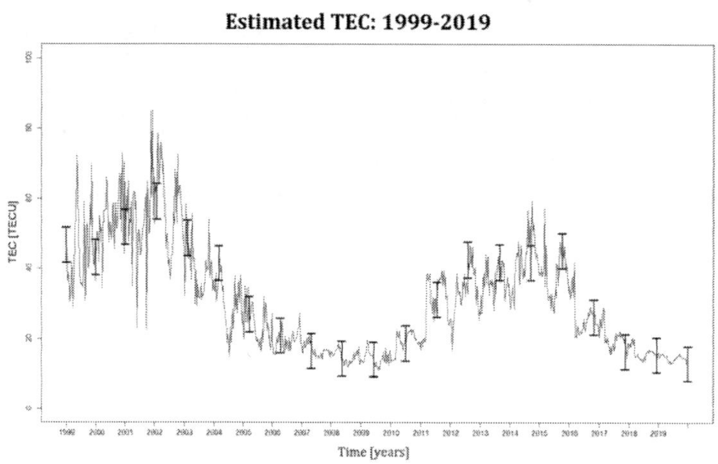

Figure 4: The calculated TEC is shown with its respective error bars. It was found after comparing with the real TEC.

The TEC values estimated and presented in the Figure 4 can be analysed as a time series from the point of view of its structural components: observed series = Trend + Seasonal effect + Residuals. This decomposition is based on elementary methods: the trend is calculated with a moving average, the seasonal effect is calculated by averaging the values of each unit of time for all the periods and then center the result. Finally, the residuals are obtained by subtracting the two previous components from the observed series.

We perform the time series analysis of the TEC series from 2009 to 2019, using the software R (version 4.0.5) a free software environment for statistical computing and graphics [16]. The following graph shows the seasonality, trend and random noise for the whole annual series.

In the graph it is observed that the series is not stationary. The series shows a trend with two maxima which pick up around 2002 and 2015 years, coinciding with the observed solar maxima of solar cycles 23 and 24. The second peak is of lower amplitude. There is a very marked seasonality (increases in the equinoxes months). Also, the amplitude of the fluctuations changes with time, so the variability is not constant either. The noise is around one, except for near the peaks where anomalies are seen. The series is quite stable.

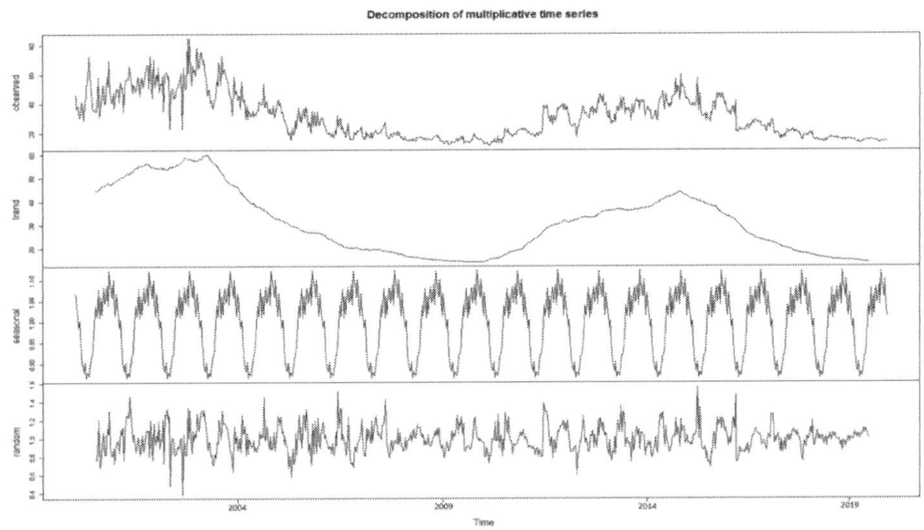

Figure 5: Time series analysis of the annual TEC from 1999 to 2019 period.

Journal of Physics: Conference Series **2238** (2022) 012010 doi:10.1088/1742-6596/2238/1/012010

6. Discussion and conclusions

The variation of X-ray, ultraviolet (UV) and of extreme ultraviolet (EUV) ionizing solar radiation is the main cause of ionospheric variability. Solar radiation produces the ions and electrons that make up most of the ionosphere by ionizing primarily atomic and molecular oxygen and nitrogen. The concentration of electrons is proportional to the intensity of the photo-ionizing radiation, which depends largely on the solar zenith angle and the composition of the neutral atmosphere. The resulting ions are redistributed by vertical diffusion, neutral winds, and by interaction with the electric and magnetic fields of the ionosphere (electrodynamics). Variations in solar ionizing radiation cause TEC to vary throughout the day, and photo-chemistry determines the regional patterns of these variations. There are spatial and temporal behaviors that deviate from those expected only with photo-chemistry. It is well known that the so-called equatorial ionospheric anomaly (EIA) involves dynamic motions of the ionospheric plasma driven by its interactions with the geomagnetic field, which in turn are driven by geographically in-homogeneous solar-generated thermospheric winds [5].

A solar flare, a sudden and explosive release of highly energetic radiation (X-rays and UV) from an active region of the Sun, immediately causes an increase in the rate of ionization and changes in the electrodynamics of the ionosphere. Consequently, there is an increase in electron density and an enhancement in ionospheric TEC that depends on the class of solar flare. These density variations differ for different altitudes and are called sudden ionospheric disturbances (SIDs). In an ongoing investigation, we study the SID in the Ecuadorian region (paper in preparation).

Here are some conclusions:

- For the first time, the estimation of the total content of electrons for Ecuador is presented, using radio signals from the GPS global positioning system. The TEC allows a better understanding of the spatial and temporal variation of the ionosphere for corrections in communication and satellite navigation systems.
- The results that were found for the total electron content in the equatorial ionosphere were satisfactory. The dispersion model used to calculate the TEC fit very well the observed values given by the IGS network. The χ^2 statistic gives a goodness of the fit about 0.88, so the fitting could be considerate acceptable.
- The fitting could be improved adding the magnetic field to the dispersion model, adding instrumental and signal detection error and processing data from further local stations located in the same territory of Ecuador.
- The daily and weekly values of TEC that we found amount to 100 TECU and the annual ones to 80 TECU, which corresponds to the well-known fact that the equatorial region is characterized by the highest values of electron density, and in this case, with peaks that are reached at noon in the Ecuadorian region.
- The time series analysis of the TEC data for the whole period of twenty-years from 1999 to 2019 clearly shows the tendency to reach peaks correlated with the maximum occurrence of solar cycles. The series is quite stable but not stationary (there is a trend) and the noise is around the baseline (around one; chi-square statistic) with anomalies close to the occurrence of the TEC peaks. So, the increase or decrease in TEC is strongly related to the solar cycle and the solar activity that it entails. Thus, for the years close to solar maximums there is an increase, while for the years close to solar minimums there is a decrease in the TEC values.
- In order that in the future the equatorial TEC study proposed in this work could be strengthened, data from several GPS receivers that are located in the same geographical

area (or nearby) could be used. In this way, the different results obtained could be averaged and thus reached a more general result that would undoubtedly be closer to the real one.

Bibliography

[1] Chih Hung Jason Chao. *Improved modelling of high precision wide area differential GPS.* PhD thesis, University of Nottingham, 1996.

[2] S Close, A Fletcher, M Dunham, and I Linscott. Coherent matched filter signal-processing algorithms for probing the ionosphere using broadband rf data. *Journal of Geophysical Research: Space Physics*, 116(A12), 2011.

[3] K Davies and GK Hartmann. Studying the ionosphere with the global positioning system. *Radio Science*, 32(4):1695–1703, 1997.

[4] Kenneth Davies. *Ionospheric radio.* Number 31. IET, 1990.

[5] J. H. Dellinger. Sudden ionospheric disturbances. *Terrestrial Magnetism and Atmospheric Electricity*, 42(1):49–53, 1937.

[6] Martin PM Hall and Leslie W Barclay. Radiowave-propagation. *NASA STI/Recon Technical Report A*, 90, 1989.

[7] Win Zaw Hein, Yoshitaka Goto, and Yoshiya Kasahara. Estimation method of ionospheric tec distribution using single frequency measurements of gps signals. *International Journal of Advanced Computer Science and Applications*, 7(12):1–6, 2016.

[8] RD Hunsucker. Radio techniques for probing the terrestrial ionosphere, phys. chem. *Space*, 22:293, 1991.

[9] Edwin C Jones. Ionospheric physics of radiowave propagation. *Dept. of Physics and Astronomy, University of Tennessee, Knoxville, TN37996*, 12, 2004.

[10] A Komjathy and RB Langley. The effect of shell height on high precision ionospheric modelling using gps. In *Proceedings of the 1996 IGS workshop international GPS service for geodynamics (IGS)*, volume 203, 1996.

[11] Attila Komjathy. *Global ionospheric total electron content mapping using the Global Positioning System.* PhD thesis, University of New Brunswick Fredericton, 1997.

[12] Attila Komjathy and Richard Langley. Improvement of a global ionospheric model to provide ionospheric range error corrections for single-frequency gps users. In *presented at the ION 52nd Annual Meeting*, volume 19, page 21, 1996.

[13] RB Langley. The effect of the ionosphere and troposphere on satellite positioning systems. In *Proceedings of Symposium on Refraction of Transatmospheric Signals in Geodesy*, volume 97. Netherlands Geodetic Commission, Publications on Geodesy, New Series, 1992.

[14] AJ Mannucci, BD Wilson, DN Yuan, CH Ho, UJ Lindqwister, and TF Runge. A global mapping technique for gps-derived ionospheric total electron content measurements. *Radio science*, 33(3):565–582, 1998.

[15] Bradford W Parkinson, Per Enge, Penina Axelrad, and James J Spilker Jr. *Global positioning system: Theory and applications, Volume II.* American Institute of Aeronautics and Astronautics, 1996.

[16] R Core Team. *R: A Language and Environment for Statistical Computing.* R Foundation for Statistical Computing, Vienna, Austria, 2021.

[17] Stefan Schaer, Gerhard Beutler, and Markus Rothacher. Mapping and predicting the ionosphere. In *Proceedings of the 1998 IGS Analysis Center Workshop, Darmstadt, February 9-11, 1998*, 1998.

[18] Stefan Schaer, Gerhard Beutler, Markus Rothacher, and Timon A Springer. Daily global ionosphere maps based on gps carrier phase data routinely produced by the code analysis center. In *Proceedings of the IGS Analysis Center Workshop 1996*, 1996.

[19] Iain Robert Webster. *A regional model for the prediction of ionospheric delay for single frequency users of the Global Positioning System.* University of New Brunswick, 1994.

[20] David Wells, Norman Beck, Alfred Kleusberg, Edward J Krakiwsky, Gerard Lachapelle, Richard B Langley, Klaus-peter Schwarz, James M Tranquilla, Petr Vanicek, and Demitris Delikaraoglou. Guide to gps positioning. In *Canadian GPS Assoc.* Citeseer, 1987.

[21] Norsuzila Ya'acob, Mardina Abdullah, and Mahamod Ismail. Gps total electron content (tec) prediction at ionosphere layer over the equatorial region. In *Trends in Telecommunications Technologies.* IntechOpen, 2010.

Effects of movement artifacts in nuclear hybrid modalities for image diagnostic

Jéssica Núñez Sánchez[1*], Pedro Escudero[2], Julio Valverde Morán[3], Julia Garayoa Roca[3], Margarita Chevalier del Río[1]

[1] Universidad Complutense de Madrid, Facultad de Ciencias Físicas, Madrid-España.
[2] SISAu Research Group, Facultad de Ingeniería y Tecnologías de la Información y la Comunicación, Universidad Tecnológica Indoamérica, Ambato, Ecuador.
[3] Fundación Jiménez Díaz Madrid, Servicio de Protección Radiológica, Madrid-España.

Corresponding author's e-mail address: jessivaleria@yahoo.com

Abstract. In the obtention of medical images, the patients' movement can modify the identification of the body components in an image. The combination of imaging techniques may not always be a solution to improve the imaging quality; therefore, an artifact analysis is commonly required prior to applying an imaging procedure in patients. In this work, we systematically evaluated the movements' artifacts caused by the patients' breathing during the images acquisition and their impact on the fusion of SPECT and CT modalities. We used a specific phantom placed on a platform to emulate the respiratory movement, finding artifacts not appreciable under the standard condition used to obtain the SPECT images due to its low spatial resolution. The artifacts produced a deformation of elements on the images. Therefore, image processing was necessary to identify the registration accuracy with SPECT and CT modalities in two states (phantom at rest and for a phantom with simulated respiratory movements). A systematic difference was obtained for the first case (11.7 mm), and a range of (7.4 mm to 16.1 mm) for the second one. For the volumes' evaluation, the optimal threshold value for CT was 0.40 and for SPECT was 0.25, giving a rapid solution to reduce the artifacts' impact on medical images.

1. Introduction

In biomedical imaging, hybrid devices integrate two imaging modalities for sequential acquisition, allowing the acquisition of merged images that are valuable tools in diagnostics and treatment. These images combine the information provided by each modality. The image fusion software uses the known positions of the scan table to create the fused image [1]. One of the most common hybrid equipment in Nuclear Medicine services is the Single-photon Emission Computed Tomography (SPECT-CT) which allows the combination of the metabolic information provided by the volumetric image generated by the gamma camera (that allows the obtention of three-dimensional representations of the radiopharmaceutical distribution in the patient), and by the anatomical information provided by the Computer Tomography (CT) images. The use of phantoms represent a common practice in imaging acquisition procedures both for parameter setting and for radiopharmaceutical dosimetry evaluation

(phantoms can be of different shapes, sometimes amorphous and with a defined form and unique characteristics, it depends of the body regions and organs on study) [2], [3].
Additionally, the organ motion produced by respiratory and cardiac movements is still a significant challenge for the accurate delivery of radiotherapy and diagnostics when using hybrid equipment [4]–[7]. Due to the differences in the acquisition time of SPECT and CT images (CT has a rapid acquisition in comparison with SPECT) the effect of the patient's respiratory movement on the two acquired imaging modalities will also be different, producing an impact on dosimetry. An example of it is the treatment of liver tumors by radioembolization [8]–[10]. In this treatment, microspheres labeled with 90Y (high-energy beta emitter) are introduced via the femoral route through a catheter. To plan this treatment, a simulation was performed with macroaggregates of albumin (MAA) labeled with Tc-99m to assess the pulmonary shunt and the uptake of the tumor and the healthy liver to be radioembolized. The patient's breathing causes a longitudinal movement of the organs of the abdominal cavity due to the displacement of the diaphragm, which in this case could affect dosimetry [8], [11].
The objective of the present work is to evaluate the effect of movement artifacts caused by the patient's breathing during the acquisition of studies, and to analyze their impact on the fusion of SPECT and CT images. For this purpose, we used a PTW L981605 phantom built according to the CEI 61675-1 specific standard for Nuclear Medicine and a QUASAR platform that simulates respiratory movement (14 breaths per minute), where commercial or self-built phantoms can be placed. The image acquisitions have been made with a filled phantom with Tc-99m (half-life 6.01 h) in different isotopes concentrations to emulate dosimetry. We systematically evaluated the movement artifacts caused by the patient's breathing during the acquisition of images and their impact on the fusion of SPECT and CT images, considering two aspects: evaluating the image register and volume segmentation through the threshold.

2. Materials and Methods

2.1. SPECT-CT Equipment

A Siemens Symbia Intevo Bold SPECT-CT Scanner of the Nuclear Medicine Service at Villalba General Hospital, Madrid was used for the image acquisition. This machine is a hybrid system that combines two imaging modalities SPECT, and CT, Figure 3A. The CT system has an X-ray tube with focus sizes of 0.8 x 0.5 mm, and 0.8 x 0.7 mm, which enables the image acquisition at 80, 110 and 130 kV. The high-definition detector array integrates 24 rows of ceramic detectors that allow the simultaneous acquisition, up to 16 slices, with 0.6 mm minimum slice thickness. In addition, the equipment includes a dose modulation system, Care Dose 4D, that modulates in real-time when the examinations are performed, allowing the regulation of mA according to the patients' characteristics. The Gamma camera gives a Field of View (FOV) of 38.7 × 53.3 cm. and includes 3 couple collimators for use according to the radioisotope (High energy(I-131), Medium Energy (Ga-67), and Low Energy High-Resolution (Tc-99). The open gantry of 70 cm allows a rotational range of 540°, speed of 0.03 to 3 rpm., and accuracy of 0.1 degrees. Furthermore, an infrared system allows real-time monitoring of the patients' contour, minimizing the distance from the patient to the collimator about to 1.2 cm; in whole-body [12].

2.2. Phantom

The phantom PTW L981605 was used to emulate the respiratory movement Figure 1. The phantom presents an asymmetric shape (anthropomorphic), which simulates the thorax and abdomen regions of the human body; it also contains hollow glass spheres that are also refillable with an internal diameter of 10, 13, 17, 22, 28, and 37 mm. Those inserts can be filled or emptied from the outside through glass capillaries with the help of a thin injection needle of similar length to the capillary, allowing easy use. The volume of the phantom is approximately 9.6 liters. The phantom spheres with different dimensions simulate hot, warm, and cold lesions. These can also determine the effect of partial volume and the spatial and contrast resolution [13].

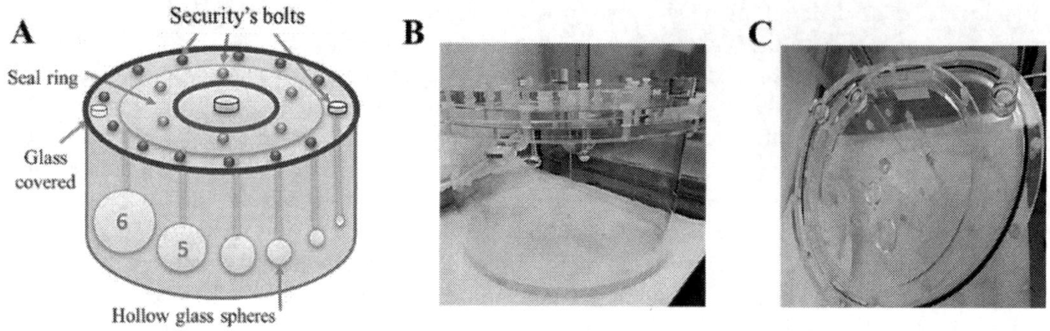

Figure 1.(**A**) Diagram of the PTW-L981605 phantom with the spheres to be filled with Tc-99m activity (spheres 5 and 6 under study) (**B**). Phantom front view (**C**) Phantom top view.

2.3. QUASAR Platform
To simulate respiratory movement, the QUASAR Platform, 35 x 35 cm was used, this platform is designed to move phantoms in a plane following different waveforms that are intended to simulate respiratory movement Figure 2C. The Respiratory Motion QA Software allows the platform movement activation importing different waveforms from patients. For this case, a sinusoidal movement pattern was chosen in the axial axis with an amplitude of ±0.5 cm. and 14 breaths per minute frequency, which corresponds to the mean displacement due to respiration of lesions located in the abdominal cavity in adults [14].

2.4. Experimental setup.
We made 6 SPECT-CT image acquisitions with the phantom filled with water and radioactive background, then at rest and in motion. Two phantom spheres were considered for analysis, those with the largest diameter and volume, the smallest phantom spheres were not correctly visualized in the SPECT images. For threshold, spheres number 5 and 6 with 11.5 and 26.5 ml in volume were used, respectively.

The phantoms' spheres were filled with radioisotope Tc-99m (half-life 6.01 h) in two different concentrations. The activity introduced in the spheres and background was calculated to produce a signal/background ratio similar to what can be found in clinical images. Table 1 shows the sequence of image acquisitions

Table 1. Sequence of image acquisitions in SPECT and CT using the phantom

Acquisition number	At rest	On movement	Without background	With background
1	x		x	
2		x	x	
*3	x			x
*4		x		x
*5	x			x
*6		x		x

* Acquisitions with movements

The parameters used to acquire images with SPECT were 64 projections of 20 s for each projection, at 140 keV, and a window size of 15%. A matrix of 128 x 128 and a FOV of 613.79 x 169.79 mm.

For CT we used a rotation time of 0.6 seconds at 130kV with a slice size of 0.75 mm. A matrix of 512 x 512, and a FOV of 500 x 500 mm. The movement produced by the QUASAR Platform describes an M.A.S. programmed with 14 breaths/minute and amplitude ± 0.5 cm.

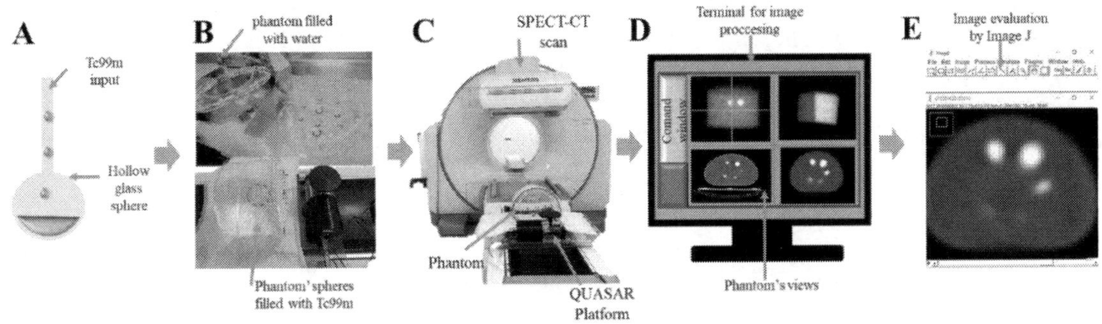

Figure 2. Experimental sequence (**A-B**) Phantom spheres filled up with water then with Tc99m (**C**) QUASAR and phantom setup placed on the SPECT-CT Unit table for image acquisition. (**D**) Image processing terminal. (**E**) Image J processing.

2.5. Evaluation of the image registration

To check the correct fusion of the SPECT and CT images, we compared the coordinates (x, y, z) of the center of the spheres. The imaging system employs an absolute reference system so that the coordinates of the spheres measured in each imaging modality are directly comparable. The comparison was done using the ImageJ software [15] and visualizing the axial, coronal, and sagittal slices with the Orthogonal Views tool. The x axis represented the left-right axis of the examination table, the y axis was perpendicular to the table's axis, and the z axis coincided with the table, Figure 3A.

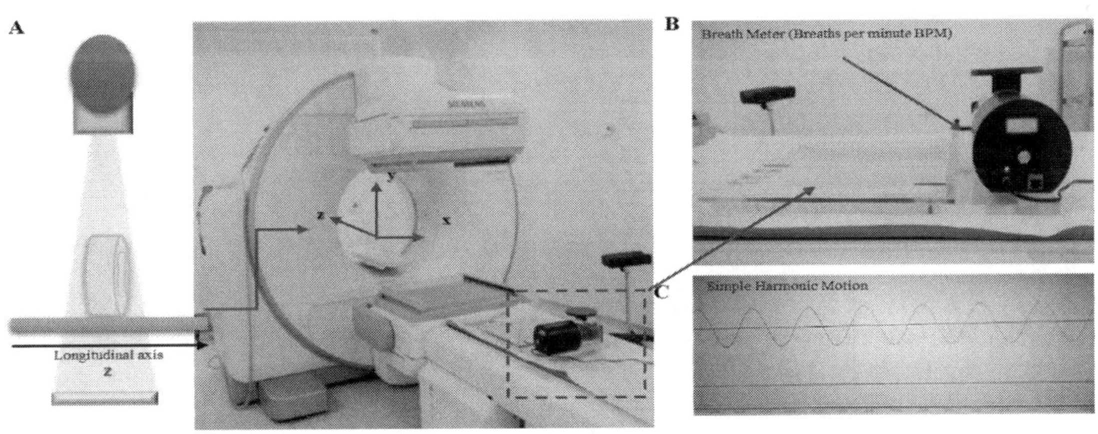

Figure 3. (**A**). Coordinate system used at work (**B**). QUASAR Platform and (**C**). Sinusoidal Graph of the M.A.S in acquisitions with movement.

For a precise center location in the SPECT images, the images were scaled to achieve the same voxel size employed on CT images. The deviation (Δx, Δy, Δz) between their centers were obtained using bilinear interpolation.

2.6. Volume Segmentation by threshold

To measure the sphere volume, a segmentation method was used by thresholding to identify the sphere location in grayscale images [16]. The optimal threshold was obtained with the phantom at rest then used to evaluate the phantom in motion. The segmentation was done using the software Image J, and considering the spheres' volume, matrix size, thickness, and range.

To find the optimal threshold value for segmentation was used the expression $thershold = t(vpmax^s - vpaverage^b) + vpaverage^b$, where $vpmax^s$ is the maximum pixel value of the sphere in all its tomographic slices, $vpaverage^b$ is the average pixel value of a Region of Interest (ROI) located in the surrounding area of the sphere, and t is a parameter defined in the range [0 1]. This parameterization has been chosen to obtain a thresholding criterion independent of the background signal level.
In the segmentation, we selected the pixels of interest by choosing the threshold values through the parameter t in order the determine the useful part of the sphere volume, Figure 4.

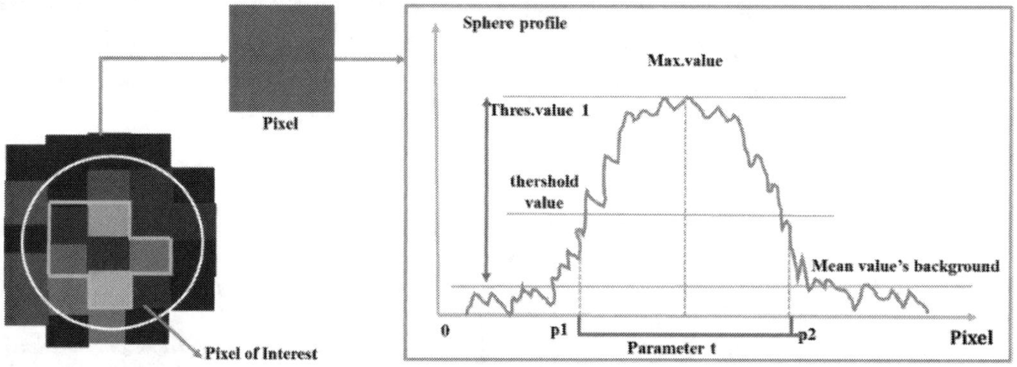

Figure 4. Interest pixels of the spheres and sphere profile elements describe the threshold equation.

The image pixels for CT contain information represented in Housnfield Units(UH), but in the case of SPECT, a pixel represents a radioactive event that captures the gamma camera detectors.
The pixel and matrix size is given by File of view (FOV) in mm, meanwhile the voxel is represented for the pixel and thickness of cutting. Equation 1 describes the relation of pixel, FOV, and matrix.

$$pixel = \frac{FOV(mm)}{matrix} \qquad voxel = pixel \times thickness\ of\ cut \qquad (1)$$

The pixel characteristics allowed us to describe two-dimensional areas, the deviation, and displacements such as the distance between the spheres, detecting artifacts measured in a range of few pixels.

3. Results

3.1. Evaluation of the image registration.

The images obtained by SPECT and CT present a small variation compared to when the phantom is at rest and in movement for x and y coordinate. However, in the z coordinate, the variations are notable for the two phantom states, visualizing an artifact associated with the displacement of the spheres for the image acquisition with CT. For this case, the spheres adopted an oval, elongated, or flattened shape depending on the coordinate observed, as shown in Figure 5. The SPECT images were scaled by bilinear interpolation to improve the low resolution.

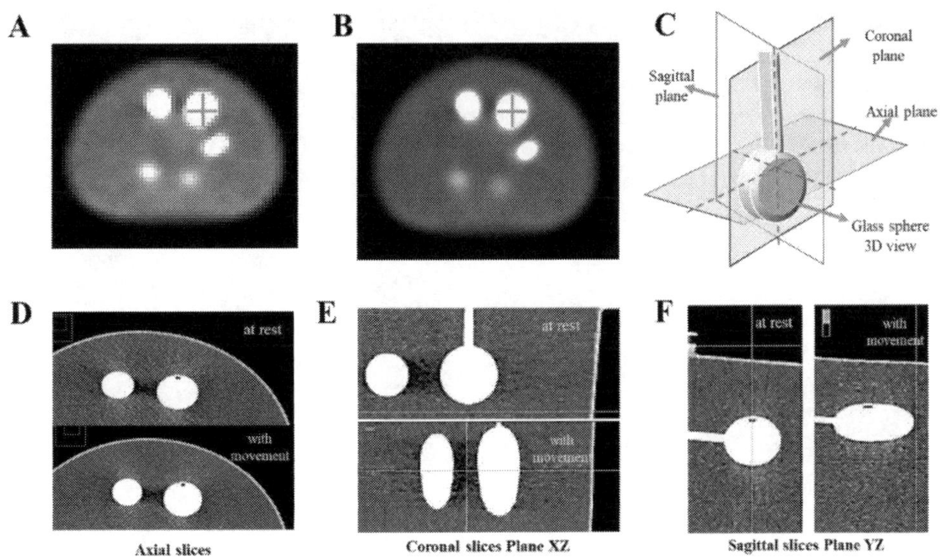

Figure 5. (**A**). SPECT image (axial view) without applying scaled (**B**) Scaled SPECT image obtaining the coordinates (x, y, z) at the center of the sphere. (**C**) Anatomic planes of the glass sphere for CT acquisitions with orthogonal views (axial, coronal, and sagittal slices), (**D**) CT Axial slices at rest and with movement (**E**), coronal slices, where the movement produces an ovoid shape in the sphere image. (**F**), sagittal slices where the movement produces the deformation of the sphere image.

The coordinate displacements of the sphere images were calculated, including the errors for the two modalities supporting that the displacements were inherent to the sphere' movement.
The phantom movement was read in z axis. Table 2 shows the systematic difference between the phantom centers; 11.7 mm on average for the phantom at rest. For the phantom in motion, the differences were between 7.4 to 16.1 mm. These differences could be due to a possible table switch when moving off the rail when the modality from SPECT to CT is produced (The equipment acquires images in two modalities sequentially. First, the SPECT image is acquired, then the table leaves the rail that is affixed to the ground and moves to the CT to acquire images).

Table 2. Coordinates (x, y, z) differences in mm for the sphere's centers in the SPECT and CT acquisitions. (S/B=Signal/Background Ratio)

x axis	**Large Sphere**			**Small sphere**		
Acquisitions Δx (mm)	S/B = infinity	S/B = 11.9	S/B = 5.5	S/B = infinity	S/B = 11.9	S/B = 5.5
At rest	-1.5	-0.5	-0.6	-0.6	0.2	0.3
With movement	-1.5	-0.7	-1.2	-0.3	0.2	-1.2
y axis						
Acquisitions Δy (mm)	S/B = infinity	S/B = 11.9	S/B = 5.5	S/B = infinity	S/B = 11.9	S/B = 5.5
At rest	-2.4	-2.9	-3.1	-3.5	-2.5	-3.5
With movement	-3.9	-2.9	-3.1	-3.5	-3.5	-3.7
z axis						
Acquisitions Δz (mm)	S/B = infinity	S/B = 11.9	S/B = 5.5	S/B = infinity	S/B = 11.9	S/B = 5.5
At rest	11.7	12.5	10.5	11.7	12.5	10.5
With movement	16.1	7.4	11.0	16.1	7.4	11.0

When the acquisition coincides with the sphere position in one of the extreme positions of the respiratory cycle, it produces that the sphere has a flattened appearance, and its position has been displaced in relation with the sphere's average position for all respiratory cycles. Alternatively, if CT acquisition occurs at an instant in the middle of the respiratory cycle, the sphere could be elongated.
Figure 6 shows two SPECT images with the phantom in motion with 14 breaths per minute under a motion range of $\pm 2\ cm$, and where the sphere elongation was visualized.

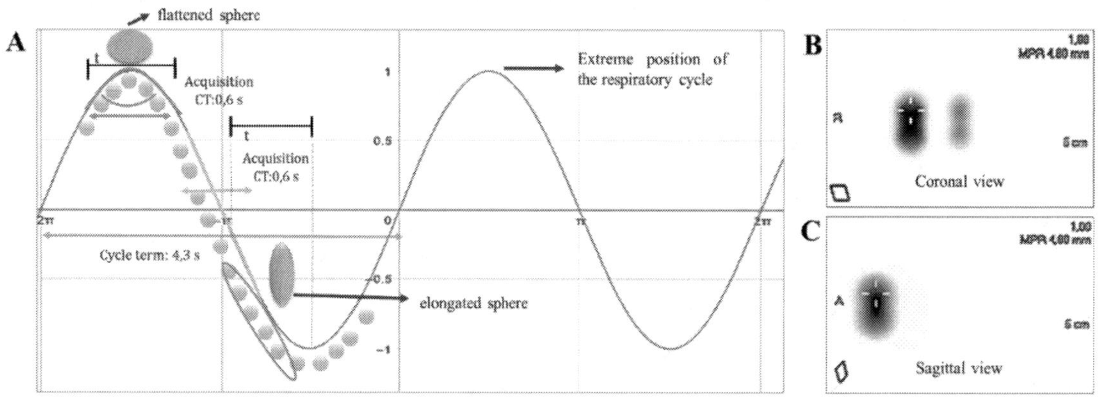

Figure 6. (**A**) M.A.S produced in the respiratory cycle and acquisitions in CT, (**B-C**) elongation of the spheres in coronal and sagittal views produced by a SPECT acquisition with movement.

3.2. Volume segmentation through a threshold.
For a range of 0.05 to 0.9 with steps of 0.05, the parameter t values were calculated to obtain the optimal threshold. Figure 7 shows the percentage deviation between the measured volume for the large and small spheres, respectively.

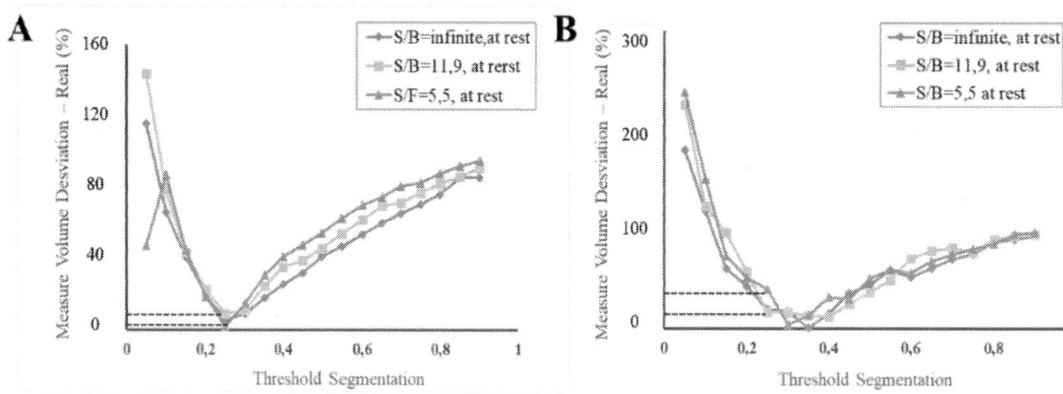

Figure 7. Segmentation threshold values for the (**A**) large sphere (**B**) small sphere

For SPECT images, the optimal threshold value was $t = 0.25$, and $t = 0.40$ for CT. The threshold value selected for segmentation in SPECT images reproduces the spheres' volume reasonably well in the acquisition at rest, with less than 20% of the original and measured volume difference. For the smaller sphere S / B radio = 11.9, the difference was 40% between the original and measured volume Table 4. Regarding acquisitions with phantom' movement, $4\ ml$ volume variation was observed compared to the phantom at rest for the large sphere S /B ratio = 5.5, which corresponds to a difference of 36 voxels. For CT images with the phantom at rest, the volume estimated was less than 10%. Moreover, CT is insensible to the radioactive background, and the estimation is independent of the ratio S/B.

Table 3. Segmentation umbral threshold for SPECT and CT. Volumes measured (ml).

SPECT (0.25)	**Large Sphere**			**Small sphere**		
Acquisitions	S/B = infinity	S/B = 11.9	S/B = 5.5	S/B = infinity	S/B = 11.9	S/B = 5.5
At rest	28	28.9	26.0	13.8	13.5	16.1
With movement	28.2	28.6	29.9	15.1	15.7	18.1
CT (0.40)	**Large Sphere**			**Small sphere**		
Acquisitions	S/B = infinity	S/B = 11.9	S/B = 5.5	S/B = infinity	S/B = 11.9	S/B = 5.5
At rest	24.4	28.9	28.9	13.8	13.5	16.1
With movement	19.8	28.5	28.6	15.1	15.7	18.1

Table 3 shows the volume values calculated using the t parameters. The volume for the large sphere was 26.5 ml, and for the small sphere, 11.5 ml.

Table 4. Threshold percentages values in SPECT and CT acquisitions

	SPECT			**CT**		
Phantom spheres	S/B = infinity	S/B = 11.9	S/B = 5.5	S/B = infinity	S/B = 11.9	S/B = 5.5
Large sphere at rest	6%	9%	2%	-8%	-10%	-8%
Small sphere at rest	20%	17%	40%	-2%	-2%	-2%

4. Discussion

4.1. Image registration

For the image evaluation was used an absolute reference system that allows us to compare the spheres centers coordinates for SPECT and CT images. In this evaluation, the deviations were less than 1.5 mm for all of them, these small distances between the x axis correspond to less than 2 pixels in CT images. Therefore, this supposes a small difference between the sphere's centers, considering that the locations of the spheres have been made subjectively. This result was expected since there was no movement on the phantom in the x direction. On the y axis, the difference between the sphere centers were -3.1 to -2.4 mm. for the large sphere and -3.7 to -3.5 mm for the small one, these differences are the same with or without the respiratory movement and correspond to a systematic error given by the sequence of image acquisition by SPECT and CT.
Meanwhile, for the z-axis, the most significant differences were observed for both imaging modalities. This is the most critical axis due to the respiratory movement and the contribution of table displacements. The difference is not attributed to the S/B ratios, as CT is insensitive to the radioactive background.

The CT acquisition time was 0.6 s due to the tube rotation. Comparing with the respiratory movement simulated on the phantom with a duration of 4.3 s per cycle (14 breaths per minute), the CT acquisition coincides with a fraction of the respiratory cycle. For the SPECT images, the acquisition time was 20 s per projection, 64 projections, slower than the respiratory movement simulated in phantom. The amplitude measured in the z axis was $\pm 0.5\ cm$. The slice thickness used in SPECT was 4.8 mm. Therefore, the system did not have enough spatial resolution to capture the artifact clearly. For greater amplitudes of respiratory movement, the elongation of the spheres could be evident in the SPECT images.

4.2. Volume's evaluation in SPECT and CT

In the case of SPECT, the optimal threshold was achieved with $t = 0.25$. With this threshold value, the measured volume for the large sphere differs by less than 9% from the real volume even when there was a radioactive background. For the small sphere, the volume determination presents a difference over the 40% for the case of S / B= 5.5. In consequence, the volume determination in the small sphere was critical due to the low spatial resolution of SPECT. For CT images, the optimal threshold was achieved with $t = 0.4$, and the deviation was 8% for the large sphere and 2% for the small sphere. There was no difference between the volumes for different S/B ratios since the HUs are not sensitive to the presence of Tc-99m. The better spatial resolution of the CT (smaller pixel size and less slice thickness) and the contribution of iodinated contrast in the spheres facilitate the threshold selection for segmentation and the estimation of spheres' volume with better accuracy.

In other studies, a segmentation was applied by the threshold for different spheres with volumes from 10 mm to 37 mm of diameter, and the reported threshold values between 40% and 45% depending on the S/B ratio, volume and sphere diameter; similar parameters were used in this work. Considering these parameters is possible to apply this methodology in clinical studies, especially in treatments that imply respiratory motion [2], [17], perfusion in cardiology studies [3], radioembolization in liver with microspheres of 90Y [18], [19]. A threshold procedure is necessary for a correct image delineation in segmentation, including the area and volume delimitation to improve the resolution for SPECT/CT studies.

5. Conclusions

In this work, we evaluated the effect of respiratory movement on hybrid images obtained with SPECT-CT equipment; we used a phantom placed in a platform to simulate the respiratory movement. An artifact related to the phantom' movement has been found in the CT images. The spheres appear flattened or elongated depending on the respiratory cycle phase. In SPECT images, an artifact associated with the movement was not visible, at least under the conditions in which the study was performed. For SPECT, we found movement amplitudes greater than the pixel size where an artifact appears associated with the respiratory movement. These results corroborate the impact of artifacts produced for respiratory movements on images. To reduce the effect of artifacts in images, we applied the segmentation of the images based on voxel thresholding value that can be useful in the planning of nuclear medicine treatments (radioembolization with Y-90, for example).

Bibliography

[1] A. Brosed, Universidad Internacional de Andalucía, and Sociedad Española de Física Médica, *Fundamentos de física médica*. Madrid: Aula Documental de Investigación, 2011.

[2] M. M. A. Dietze, B. Kunnen, M. G. E. H. Lam, and H. W. A. M. de Jong, "Interventional respiratory motion compensation by simultaneous fluoroscopic and nuclear imaging: a phantom study," *Phys. Med. Biol.*, vol. 66, no. 6, p. 065001, Mar. 2021, doi: 10.1088/1361-6560/abe556.

[3] Y.-W. Yang, J.-C. Chen, X. He, S.-J. Wang, and B. M. W. Tsui, "Evaluation of Respiratory Motion Effect on Defect Detection in Myocardial Perfusion SPECT: A Simulation Study," *IEEE Trans Nucl Sci*, vol. 56, no. 3, pp. 671–676, Jun. 2009, doi: 10.1109/TNS.2009.2015446.

[4] L. Beaton, S. Bandula, M. N. Gaze, and R. A. Sharma, "How rapid advances in imaging are defining the future of precision radiation oncology," *Br J Cancer*, vol. 120, no. 8, pp. 779–790, Apr. 2019, doi: 10.1038/s41416-019-0412-y.

[5] F. Cuccia *et al.*, "Patient positioning and immobilization procedures for hybrid MR-Linac systems," *Radiation Oncology*, vol. 16, no. 1, p. 183, Sep. 2021, doi: 10.1186/s13014-021-01910-6.

[6] J. Bertholet *et al.*, "Real-time intrafraction motion monitoring in external beam radiotherapy," *Phys Med Biol*, vol. 64, no. 15, p. 15TR01, Aug. 2019, doi: 10.1088/1361-6560/ab2ba8.

[7] O. Israel *et al.*, "Two decades of SPECT/CT – the coming of age of a technology: An updated review of literature evidence," *Eur J Nucl Med Mol Imaging*, vol. 46, no. 10, pp. 1990–2012, Sep. 2019, doi: 10.1007/s00259-019-04404-6.

[8] R. Bastiaannet, M. A. Viergever, and H. W. A. M. de Jong, "Impact of respiratory motion and acquisition settings on SPECT liver dosimetry for radioembolization," *Med Phys*, vol. 44, no. 10, pp. 5270–5279, Oct. 2017, doi: 10.1002/mp.12483.

[9] A. K.-T. Tong, W. Y. Tham, C. W. Too, D. W.-M. Tai, P. K.-H. Chow, and D. C.-E. Ng, "Molecular Imaging and Therapy of Liver Tumors," *Seminars in Nuclear Medicine*, vol. 50, no. 5, pp. 419–433, Sep. 2020, doi: 10.1053/j.semnuclmed.2020.04.004.

[10] R. Bastiaannet, S. C. Kappadath, B. Kunnen, A. J. A. T. Braat, M. G. E. H. Lam, and H. W. A. M. de Jong, "The physics of radioembolization," *EJNMMI Phys*, vol. 5, p. 22, Nov. 2018, doi: 10.1186/s40658-018-0221-z.

[11] C. H. Cabrerizo, P. C. Tejero, E. C. Silva, J. P. Moreno, A. S. Hierro, and P. F. Letón, "Contorneo de volúmenes tumorales mediante segmentación por umbrales en imágenes PET: influencia de la relación señal-fondo, tamaño y movimiento de la lesión," *Revista de Física Médica*, vol. 11, no. 2, Art. no. 2, Oct. 2010, Accessed: Sep. 23, 2021. [Online]. Available: https://revistadefisicamedica.es/index.php/rfm/article/view/97

[12] S. Electromédica, "SISTEMA SPECT/CT-SYMBIA Intevo Bold," 2018. https://www.siemens-healthineers.com/es/molecular-imaging/xspect/symbia-intevo-bold

[13] P. USA, "Maniquíes para PET y SPECT L981602, L981605, L981606 y L981607," 2019. http://www.ptw-usa.com

[14] M. QA, "QUASAR™ Respiratory Motion Platform," 2019. https://modusqa.com/products/quasar-respiratory-motion-platform/

[15] W. Rasband, "National Institutes of Health," *Image J*, 2009. https://imagej.nih.gov/ij/

[16] C. Huerga Cabrerizo and P. Castro Tejero, "Contorneo de volúmenes tumorales mediante segmentación por umbrales en imágenes PET: influencia de la relación señal-fondo, tamaño y movimiento de la lesión," Madrid, 2, 2010.

[17] K. Okuda, K. Nakajima, A. Kikuchi, M. Onoguchi, and M. Hashimoto, "Cardiac and Respiratory Motion-induced Artifact in Myocardial Perfusion SPECT," *Annals of Nuclear Cardiology*, vol. 3, no. 1, pp. 88–93, 2017, doi: 10.17996/anc.17-00005.

[18] B. Kunnen, S. van der Velden, R. Bastiaannet, M. G. E. H. Lam, M. A. Viergever, and H. W. A. M. de Jong, "Radioembolization lung shunt estimation based on a 90Y pretreatment procedure: A phantom study," *Medical Physics*, vol. 45, no. 10, pp. 4744–4753, 2018, doi: 10.1002/mp.13168.

[19] W. Badar *et al.*, "Safety and efficacy of repeat Y90 radioembolization to the same hepatic arterial territory," *BJR*, vol. 94, no. 1119, p. 20200752, Mar. 2021, doi: 10.1259/bjr.20200752.

Design and construction of a low-cost 3D-printed portable LIBS system

O Ormachea[1], A Villazon[2] and O Oporto[1]

[1] Centro de Investigaciones Ópticas y Energías, Universidad Privada Boliviana, Cochabamba, Bolivia

[2] Centro de Investigaciones en Nuevas Tecnologías Informáticas, Universidad Privada Boliviana, Cochabamba, Bolivia

E-mail: oormachea@upb.edu

Abstract. Laser-Induced Breakdown Spectroscopy (LIBS) is a widely used non-destructive analysis technique, which is however costly. In this paper we present the development of a low-cost 3D-printed portable LIBS system designed and built with a Technology Readiness Level 5 (TRL 5), i.e., technology validated in a relevant environment. The proposed LIBS system is compact and includes a passive Q-switch Nd^{3+}:YAG pulsed laser, a handheld measuring device, a power supply, an optical spectrometer and a processing software. The electronic control system guarantees a stable laser triggering, thanks to the use of optical connections (optocouplers) avoiding wired electrical connections of the circuits. In addition, we found the correct energy characteristics of the passive Q-switch regime, to generate a single laser pulse, thus making the system operate more efficiently. Our low-cost 3D-printed portable LIBS system was validated with several tests with real samples of atomic elements (i.e., lithium, copper, silver, gold and lead).

1. Introduction

Analysis methods based on Atomic Emission Spectroscopy have been used for a long time and have evolved over the years towards different applications, while keeping the same basic concept: Atoms emit characteristic spectral lines when excited by some energy source; if such spectral lines can be observed in any sample, this confirms the presence of the atom even though the sample contains other elements. Also, the measured intensity of the spectral lines provides quantitative information of the element in the sample [1]. Atomic Emission Spectroscopy uses an energy source to convert a portion of matter into plasma and forces its atoms to emit photons. Spectrometers analyze the emitted radiation in order to determine the atomic composition of the sample.

The main component of the Atomic Emission Spectroscopy is the energy source used for sample excitation. Some of the most commonly used atomic emission excitation sources are the *spark electrode* and the *inductively coupled plasma* (ICP), which reach temperatures of 50000 K and 10000 K, respectively [2].

With the development of laser technology in the early 60s, several researchers investigated its use in atomic emission spectroscopy [3], and rapidly lasers became an important excitation source for atomic emission analysis [4]. In 1963, Laser-Induced Breakdown Spectroscopy - LIBS [5] introduced the analytical use of plasma generated by laser radiation on surfaces. Currently, several types of pulsed lasers are used in LIBS systems, with the Nd^{3+}:YAG type being the most common. The temperature of the laser-generated plasma varies between 8000 K and 12000 K [2].

LIBS technique has shown several advantages compared to other methods (e.g., Atomic Absorption Spectroscopy - AAS or Inductively Coupled Plasma Mass Spectrometry - ICP-MS), such as its non-destructive analysis capability, in-situ measurement capability and portability. The non-destructive analysis capability allows the LIBS technique to be used in hostile and hazardous environments for humans, and its in-situ measurement capability and portability allows it to be used in the field, thus reducing analysis time [6].

The applications of LIBS covers diverse areas, including agriculture and food [7], explosives detection [8], analysis of terrestrial and extraterrestrial minerals and rocks [9], nuclear fusion technologies [10], industrial applications [11], archaeology [12], human and animal health [13], among many others.

Although LIBS has important advantages over other atomic spectral analysis techniques, it is extremely expensive. In this sense, our long-term research effort focuses on the development of low-cost spectrometers and pulsed lasers, integrated into a functional portable LIBS system.

In this paper, we present the development of a low-cost 3D-printed portable LIBS system and a control software for in-situ and real-time detection of atomic samples, i.e., an equipment with a Technology Readiness Level 5 (TRL 5) [14]. This work is based on previous research results in the development of a Nd^{3+}:YAG pulsed lasers [15] and low-cost optical spectrometers [16,17,18] which had TRL 4 category (i.e., technology validated in a laboratory). The contribution of this paper, compared to previous work, can be summarized as follows:

- The optimization of the control electronic system to avoid pulse losses during laser triggering.
- Energy control of the optical pumping to ensure a single pulse in passive Q-switch regime.
- The design and construction of a robust 3D-printed and portable LIBS system.

The rest of the paper is structured as follows: Section 2 describes the mechanical design and 3D printing. Section 3 presents the design and development of the electronic power supply of the laser and the processing software. In Section 4, we describe the results obtained with real samples and Section 5 concludes the paper.

2. Mechanical design and 3D printing

Our proposed LIBS system consists of (a) a high-power pulsed laser that generates the plasma in the sample, (b) an optical spectrometer, and (c) a light collection system that collects the light emitted by the plasma and sends it through an optical fiber to the spectrometer. The spectrometer is responsible for the analysis of the different wavelengths produced by the plasma light.

In order to allow our device to operate in real-world field conditions, we need to ensure a robust and resistant mechanical structure, so as to keep the internal components fixed. In addition, those internal components, i.e, the electronics, optical (pulsed laser and signal collection system) and spectrometric (optical spectrometer), must be designed to be as compact as possible. Our approach is to use robust and low-cost 3D printing material, which eases the design and manufacturing of the portable LIBS prototype.

2.1. Design of the handheld measuring device

The LIBS handheld measuring device has two main functions: (a) it contains the pulsed Nd^{3+}:YAG laser and focalizes the laser radiation into the sample, and (b) it contains an optical system for collecting the light generated by the plasma that is carried to the spectrometer through an optical fiber. This mechanical part must be as ergonomic as possible in order to allow the user to simply approach and point the sample to be analyzed, without requiring any special alignment, and trigger the laser by pushing a button. This solution allows efficient collection of the light of the plasma in non-homogeneous surfaces.

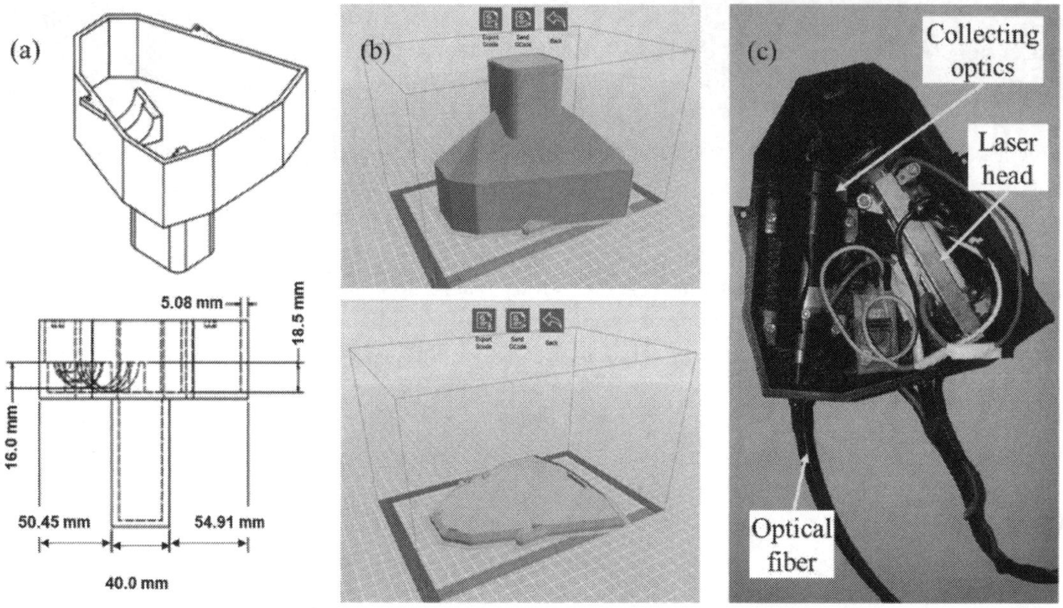

Figure 1. Mechanical design of the handheld structure (a), SolidWorks design (b) and the mounted handheld measuring device (c).

Figure 1(a) shows the design of the robust mechanical structure of the handheld measuring device, Figure 1(b) shows the 3D design in SolidWorks ready to be manufactured, and Figure 1(c) shows the assembled handheld measuring device. The mechanical structure was 3D-printed using black PLA filament to avoid, as much as possible, laser beam reflections for eye safety reasons.

The excitation light source is a small Nd^{3+}:YAG laser head which emits laser radiation at 1064 nm. It includes a flash lamp with external trigger, a YAG rod, a passive Q-switch crystal, and two mirrors, all assembled in a 25x25x105 mm compact metal case. The energy of each pulse in the passive Q-switch regime is of 20 mJ with a duration of 40 ns (see Figure 2(a)), with an energy input of 16 J at 100 μs flash lamp pulse duration. The measurement of the pulse duration was made with a high speed photodetector (11HSP-FS1 with a wavelength range of 400-1100 nm) and a 200 MHz bandwidth Tektronix oscilloscope. For the energy monopulse measurement, we used a portable laser power and energy meter (StarLite of Ophir). The maximum trigger frequency of the monopulse emission in the Q-switch regime was 0.14 Hz. This frequency was used because the laser system is only cooled by convection with the ambient air.

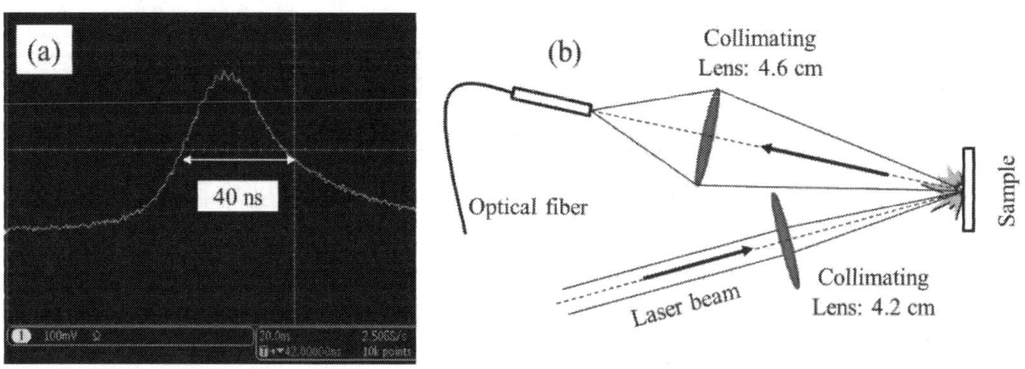

Figure 2. Oscillogram of the temporal duration of the laser pulse in the passive Q-switch regime (a) and the optical light collecting scheme (b).

To generate the plasma on the sample, the laser beam is focalized with a collimating lens (focus length: 4.2 cm). Figure 2(b) shows the optical scheme for collecting the light generated by the plasma, which is composed by a second collimating lens (focus length: 4.6 cm) and an optical fiber. The collimating lens is responsible for collecting all the light from the plasma, which is transmitted to the spectrometer through the optical fiber.

2.2. *Design of the mechanical structure of the optical spectrometer*

Based on previous work [19], we designed an optical spectrometer with a reflection configuration. It uses two compact optical objective lenses, a diffraction grating of 1200 lines/mm, and a digital monochromatic camera (Chameleon CCD digital camera) as opto-electronic sensor. The spectral resolution (measured with the method described in [16]) of the developed optical spectrometer is 0.27 nm and a spectral window of 76.9 nm.

Figure 3(a) shows the design of the mechanical structure of the optical spectrometer, Figure 3(b) shows the designs in SolidWorks, and Figure 3(c) shows the assembled optical spectrometer. A variable slit was additionally mounted on this structure to control the light coming from the optical fiber that carries the signal produced by the plasma generated in the sample. Similarly, a 3D-printed structure was built to hold the laser power supply and the electronic control circuit.

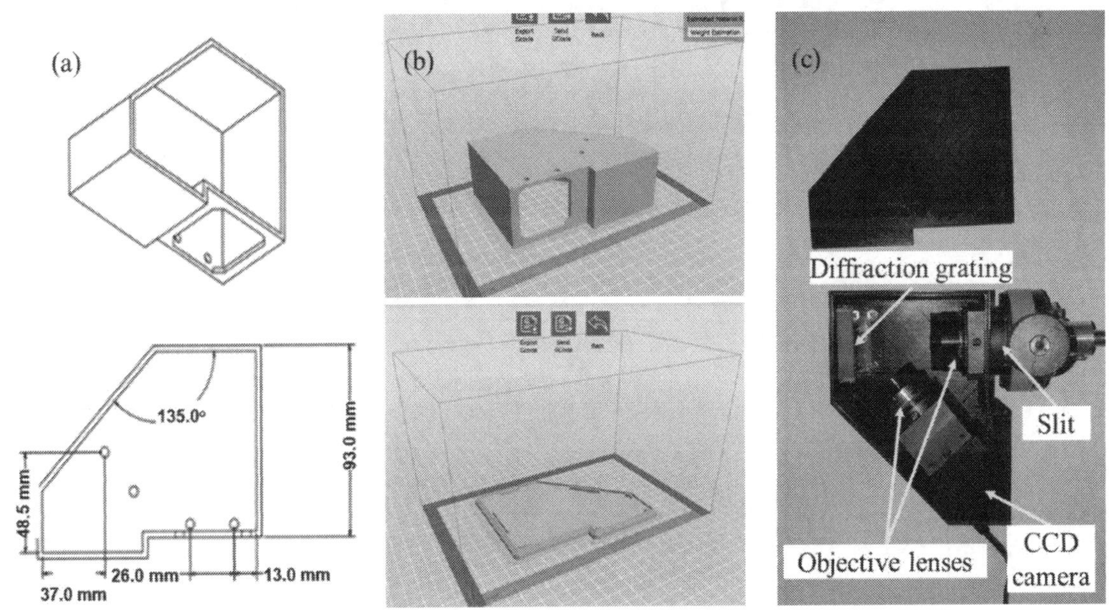

Figure 3. Mechanical design of the optical spectrometer structure (a), SolidWorks design (b) and the mounted optical spectrometer (c).

3. Electronic power supply and processing software

The electronic power supply and control circuit of the pulsed laser was fully designed and built using commercially available electronics, thus reducing the costs. In addition, we developed a customized software for the acquisition and processing of the optical signals produced by the plasma.

3.1. *Electronic power supply*

In previous work [20], we developed an electronic system for power supply and control of the laser head, which had a limited rate of successful electronic signal triggering a laser pulse (of around 60%). Furthermore, the laser worked in a passive Q-switch regime that generated several laser pulses for

each successful trigger (i.e., the total laser energy was distributed between those pulses, thus reducing the effectiveness of the LIBS). A Cr^{4+}:YAG saturable crystal was used for the passive Q-switch regime. In this paper, we optimized the power supply electronics by gradually increasing the energy applied to the laser until the optimal threshold was reached (545 VDC, energy input of 16 J) for the optical pumping. This allows to obtain a single laser pulse by controlling the energy parameters that are sent to the laser head, while keeping the passive Q-switch regime.
The maximum input energy used for pumping the flash lamp was 28 J at 720 VDC (see the oscillogram in Figure 4(a)), resulting in 4 pulses at 150 μs flash lamp pulse duration. By decreasing the input energy of the flash lamp, we obtained a reduction in the number of pulses observed in this regime. Figure 4(b) shows the oscillogram of the obtained monopulse with an energy input of 16 J.

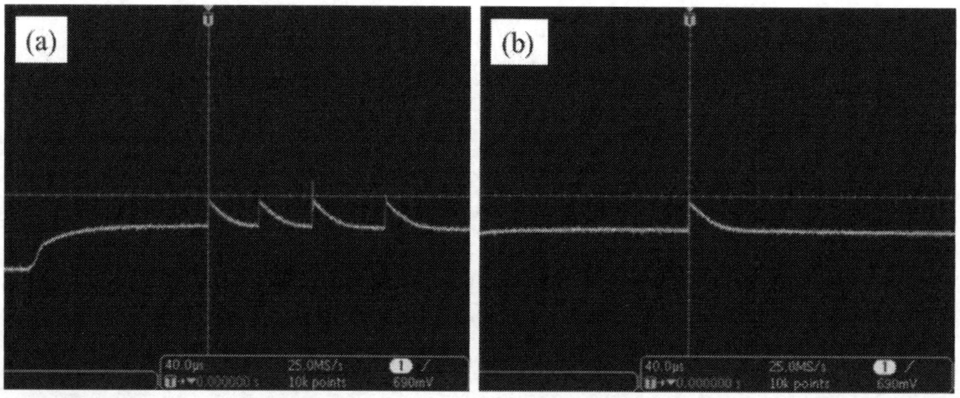

Figure 4. Oscillogram of the laser pulses in the passive Q-switch regime: four pulses are obtained with an energy input of 28 J (a), and a monopulse is obtained with an energy input of 16 J (b).

The new power supply electronics has three main components: (a) the power circuit, (b) the triggering circuit, and (c) the control circuit. The electronic circuits (a) and (b) were designed in the Proteus[1] (see Figure 5), and the corresponding PCBs (printed circuit boards) were manufactured on a CNC (computer numerical control). Figure 6 shows the assembled power circuit. In contrast, the control circuit (c) uses an Arduino Uno microcontroller which controls the stages of operation of the power circuit (a) and the triggering circuit (b).
Additionally, in order to ensure a higher success rate of the laser pulse triggering (from around 60% to 99%), we apply the following optimizations in the electronic design:

- We avoid wired electrical connections between the three main aforementioned circuits of the power supply electronics, and use optical connections instead (optocouplers), in order to eliminate any electrical noise and high-voltage interference between the circuits.
- We use capacitors to store the electrical energy in the power circuit, required to activate the laser head. These capacitors are charged only when the laser is triggered, instead of a continuous charge of our previous design [20].

[1] https://www.labcenter.com/pcb/

Figure 5. Design of the electronic circuit in the Proteus PCB Design software.

Figure 6. The assembled power circuit

All the circuits of the LIBS electronic system were assembled and inserted into a compact 3D-printed case.

3.2. *Processing software*

We developed a custom software for the acquisition and processing of the spectrum produced by the analyzed samples. The software also controls the Chameleon CCD digital camera and constantly captures the light images (every 55 ms) after the laser radiation generates plasma in the sample. The software detects any drastic change in the intensity of the captured light, and stores the images. Then, the software processes the captured image and generates the corresponding spectral lines. Note that the Chameleon CCD digital camera used in the device, has a minimum trigger delay time of 0.1 ms, which is too high compared to the ideal delay of 500 ns, to capture the image after the generation of plasma [2]. This ideal delay is critical for quantification of atomic elements, however, since our current work focuses on identification of atomic elements, the proposed solution based on constant image capture is adapted to this purpose. Adding a fast triggering electronic control system to synchronize the laser pulse with the image capture may increase the cost of the device. Such a development will be covered in future work.

The developed software has an embedded database with reference spectral information (peak values of the main spectral lines) of several elements of interest from the periodic table. These references were obtained from the National Institute of Standards and Technology (NIST)[2]. Our software compares this database with the spectra obtained by the proposed LIBS system, by applying a peak detection algorithm. This allows qualitatively identifying the existence of the element of interest in the analyzed sample and a visual comparison between the generated spectra and the references.

Figure 7 shows the main window of the control panel of our LIBS software. The software allows users to observe the images captured from the camera in real-time. It also allows the calibration of the spectrometer as a function of the wavelength. Finally, it is possible to select elements of the database

[2] https://www.nist.gov/pml/atomic-spectra-database/

and perform the processing of the images to obtain the spectral lines of the analyzed samples and compare them with the referential data.

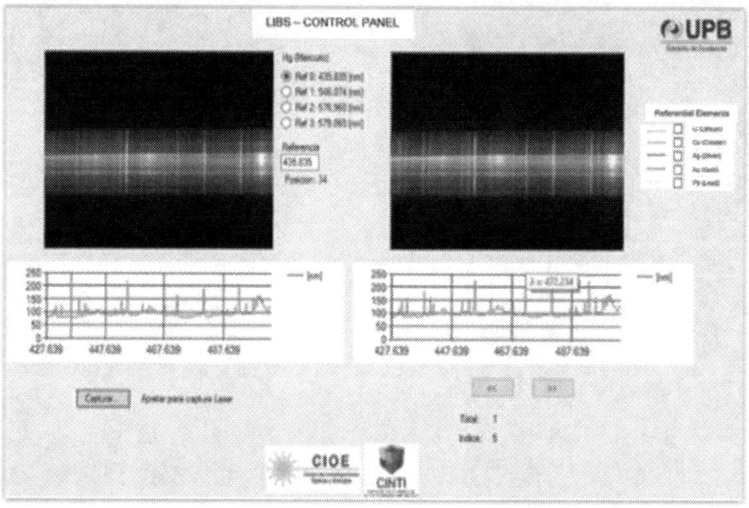

Figure 7. The LIBS control software.

4. Results

The proposed compact and low-cost 3D-printed portable LIBS system (shown in Figure 8) includes an optical spectrometer, an electronic power supply with a control of the Nd^{3+}:YAG pulsed laser, and a handheld measuring device that collects the light produced by the plasma on the sample and sends it through an optical fiber to the spectrometer. The dimensions of the different components of the device are: 14.5 cm X 15.4 cm X 12 cm for the handheld measuring system, 16.8 cm X 11.8 cm X 11.5 cm for the power supply case, and 13.2 cm X 10.4 X 9.60 cm for the spectrometer.

Thanks to the improvements made in electronics, mechanical design, robustness and portability, our LIBS system becomes a Technology Readiness Level 5 (TRL 5) equipment [14]. The TRL 5 category confers the status of functional equipment for a real environment. This feature is very important to be able to use our system in different application areas of the LIBS technique that require non-destructive, real-time and in-situ analysis.

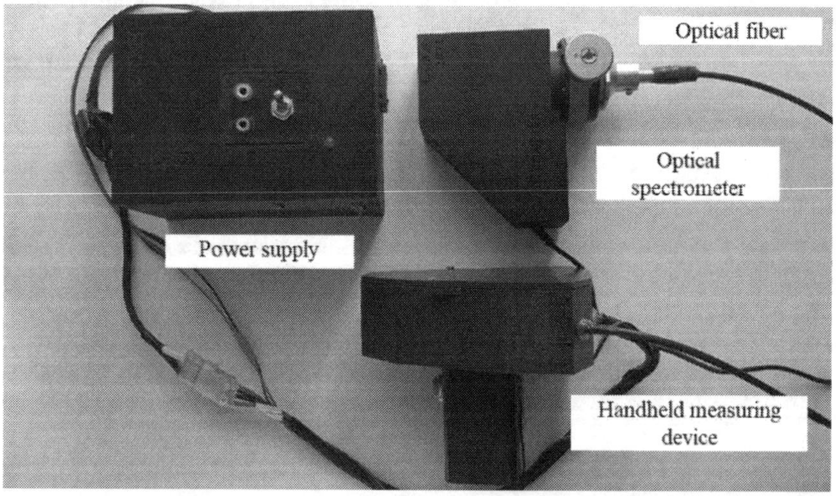

Figure 8. The complete proposed 3D-printed portable LIBS system.

To validate the proposed LIBS system, we carried out several tests with samples of atomic elements (i.e., copper, lithium, silver, gold and lead). Figure 9 shows spectral lines of copper and lithium where laser shots were made with the handheld measuring device in the real samples.
In the control software, we select the atomic element to be analyzed (i.e., Ag, Li) from the database and the corresponding reference lines are displayed over the obtained spectrum. The software compares the peaks of the obtained spectrum against the reference lines from the database (i.e. wavelengths λ_1=521.06 nm and λ_2=546.70 nm for silver; and λ_1=610.50 nm for Li). Figures 9(a) and 9(b) clearly show the coincidence of the spectral lines obtained with our LIBS system with those of the reference elements.

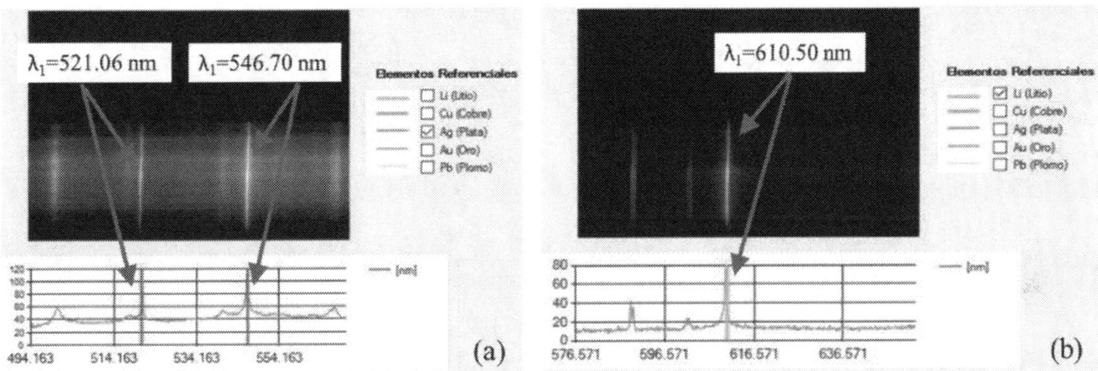

Figure 9. Spectral lines of real samples of silver (a) and the spectral line of lithium (b).

The experimental results with real samples demonstrate the validity of the proposed LIBS system for the identification of atomic elements of interest in real-time.

5. Conclusions

In this paper we present the development of a compact low-cost 3D-printed portable LIBS system, for the identification of atomic elements in-situ and in real-time. This development allowed us to improve our LIBS technology, evolving from a TRL 4 (laboratory validation) to a TRL 5 (validation in relevant environment) category. This step was possible thanks to: (a) the optimization of the control electronic system to avoid pulse losses during laser triggering, thanks to the use of optical connections (optocouplers) avoiding wired electrical connections of the circuits to eliminate any electrical noise, (b) efficient single pulse generation of the passive Q-switch regime though an energy control of the optical pumping, and (c) a robust and portable 3D-printed system, validated with real samples of different elements in a relevant environment. The estimated cost of the proposed LIBS device is around 2000 USD.
With this successful experience, our plan for future work is to reach TRL 6, i.e., demonstrate our technology in relevant environments, such as the mining industry or archeology. Also in this stage, an electronic trigger control system will be implemented to synchronize the laser pulse and the capture of the optical signals.

References

[1] Singh J P and Thakur S N 2007 *Laser-Induced Breakdown Spectroscopy* (Amsterdam: North-Holland/American Elsevier)
[2] Cremers D A and Radziemski L J 2013 *Handbook of Laser-Induced Breakdown Spectroscopy* (John Wiley & Sons, Ltd)
[3] Radziemski L J 2002 *Acta B At. Spectrosc.* **57(7)** 1109-1113

[4] Brech F and Cross L 1962 *Appl. Spectrosc.* **16** 59-64
[5] Debras-Guedon J, Liodec N and Vilnat J 1965 *J. Appl. Math. and Phys.* **16** 155–156
[6] Senesi G, Harmon R and Hark R 2021 *Acta B At. Spectrosc.* **175** 106013
[7] Nicolodelli G, Cabral J, Menegatti C, Marangoni B and Senesi G 2019 *Part I. Soils and fertilizers, Trends Anal. Chem* **115** 70–82
[8] Gottfried J, De Lucia F, Munson C and Miziolek A 2009 *Anal. Bioanal. Chem.* **395** 283–300
[9] Senesi G 2014 *Earth Sci. Rev.* **139** 231–267
[10] Li C, Feng C, Oderji H, Luo G, and Ding H, 2016 *Front. Physiol.* **11** 16
[11] Noll R, Fricke-Begemann C, Connemann S, Meinhardt C and Sturm V 2018 *J. Anal. At. Spectrom.* **33** 945–956
[12] Botto A, Campanella B, Legnaioli S, Lezzerini M, Lorenzetti G, Pagnotta S, Poggialini F and Palleschi V 2019 *J. Anal. At. Spectrom.* **34** 81–103
[13] Gaudiuso R, Melikechi N, Abdel-Salam Z, Harith M, Palleschi V, MottoRos V and Busser B 2019 *Acta B At. Spectrosc.* **152** 123–148
[14] Héder M 2017 *Innov. J.: The Public Sector Innov. J.* **22(2)** 1-23
[15] Ormachea O 2007 *Revista Boliviana de Física* **13(13)** 54-57
[16] Ormachea O, Urquidi O and Cisneros J 2013 *Revista Cubana de Física* **30(2)** 93-98
[17] Ormachea O, Villazon A and Escalera R 2017 *Opt. Pura Apl.* **50(3)** 239-249
[18] Ormachea O, Urquidi O and Casazola D 2013 *Proc. of SPIE (Porto)* vol 8785 (SPIE) 87851D87851D-8
[19] Terceros I, Ormachea O, Villazon A and Urquidi O 2016 *Revista Politécnica* **37(1)** 113-119
[20] Ormachea O, Villazon A and Terceros I 2018 *Investigación & Desarrollo* **18(1)** 81-90

Dust adhesion physics on photovoltaic surfaces and its application on assisted natural cleaning of solar trackers

G A Cruz and P Vorobiev[1]

CIMAV/Monterrey, Alianza Norte 202, PIIT, Apodaca 66628, N.L., México.

[1]E-mail: pavel.vorobiev@cimav.edu.mx

Abstract. Soiling is a problem that affects energy production of solar panels. Dust particles are deposited on the surface of the panel by the wind currents. Adhesion occurs due to Van der Waals forces, depending on the contact area of the tempered glass surface with the dust particle, due to gravitational force acting on the mass of the particles, and due to capillary forces, which are increased with the amount of humidity, depending on the climate. In this project, the positioning capabilities of a solar tracker were used to achieve better assisted natural washing under conditions of weak and heavy rain, using a "passive cleaning" when the rain is intense and an "active cleaning" when the rain is weak. For passive cleaning, the effect of varying the inclination angle while washing and drying was studied and quantified, analysing the formation of stains generated by raindrops drying. It was evident that the higher the angle of inclination the greater the effect of gravity on the drops, facilitating their sliding and falling off the panel, however, it was found that it is possible to carry out passive cleaning at an angle of inclination that does not require prominent deviation from the angle that fits the position of the sun, in fact, it is enough with an inclination angle of 30° to achieve an effective passive cleaning process. In active cleaning, a rocking movement is performed by varying the angle of inclination of the solar tracker with values of + θ and -θ, which increases the contact time of the raindrops with the surface of the solar panel, allowing a better sweep of dust particles and therefore a more complete washing even in light rain conditions. Active cleaning improves the results in surface cleaning in almost 57% when compared to passive cleaning, in light rain conditions.

1. Introduction

The increase in photovoltaic installations around the world has brought with it an increase in the investigation of the new challenges faced by the electrical energy industry produced by solar panels. When a solar panel is installed and evaluated outside of a laboratory environment, we can appreciate factors that reduce their energy production efficiency.

1.1 Soiling in solar panels

Many factors determine the optimal performance of photovoltaic modules. Climate is one of the factors that directly affects the performance of photovoltaic cells. An effect that notably affects the performance of photovoltaic modules is the so-called "Soiling", which describes a shading effect caused by air pollution, cataloged as "Soft Soiling", or by the accumulation of dust on the surface of solar panels, which is classified as "Hard Soiling", the latter has been the subject of a large number of studies that analyze its impact on the efficiency of photovoltaic modules or propose solutions to this problem [1,2].

Studies have shown that soiling can cause efficiency losses in energy production between 10 and 70% [3].

Among the important factors in the soiling process are the roughness of the surface. Rough surfaces favor the deposition of dust while a smooth surface has less risk of dust accumulation. Panel position and wind direction are also important in the fouling process. A surface closer to the horizontal position is more susceptible to dust accumulation. Strong winds can favor the cleaning of the panels, while slower winds lead to greater deposition of dust. While the speed of the wind increases the pressure on the surface of the panel is lower, which decreases the deposition of dust and vice versa. Other dust properties such as size, shape and weight play an important role in the light scattering generated by the dust and therefore in the effect it will have on the efficiency of photovoltaic devices.

Among other relevant studies on soiling, A. Kimber et. al. study the effect of soiling in California USA, concluding in his research that at least 20mm (millimeters per square meter, meaning a volume of 20 liters of rain) of rainwater is required to favorably clean the surface of a solar panel, weaker than 20mm rains can have a foundation effect worsening the soiling effect [4]. Although in Kimber´s study the energy production losses due to dirt do not exceed 5%, studies such as that of Al Soleimani et. al., carried out in several areas of Iraq, show photovoltaic devices output power losses of up to 60% varying with the seasons of the year and also verifying that the fouling of the panels is strongly dependent on the geographical position in which they are located [5].

1.2 Dust adhesion mechanisms

The dust particles that reach the surface of the panel are adhered to it by various mechanisms. Among the main ones, there are the Van der Waals forces (F_{VdW}) which consist of interactions of attraction or repulsion between atoms or molecules, it is dependent on distance and it is not a result off electronic bonds. Normally, in particle adhesion studies, Van der Waals forces are considered important when the particle has a radius less than 50 μm. Equation 1 is used for the calculation of the F_{VdW}. The gravitational force (F_g) also contributes to the adhesion forces, as shown in equation 2. It is important to highlight that at microscale the contribution of this adhesion force is small compared to the other adhesion forces. The capillary force (F_{st}) is due to surface tension, which is present when there is a layer of water between the particles and is calculated with equation 3 [6].

$$F_{VdW} = \frac{AR}{6r^2} \quad (1)$$

Van der Waals forces on a surface, A represents the material dependent Hamaker interaction constant, R is the radius of the particle, and r is the distance between the particle and the surface.

$$F_g = mpg = \frac{4}{3}\pi R^3 \rho g \quad (2)$$

Gravitational force of a dust particle on a surface, mp is the mass of the particle, g is the acceleration in free fall, and ρ is the density of the particle. The volume of the particle $V = \frac{4}{3}\pi R^3$ is given for the case of a spherical particle.

$$F_{st} = (4\pi R\gamma)\cos\theta \quad (3)$$

Capillary force due to surface tension, γ is the surface tension of the liquid, θ is the contact angle, and R is the radius of the particle.

1.3 Solar tracking

The maximum production of photovoltaic energy is obtained when the maximum amount of irradiation is collected, this is achieved by keeping the luminous flux perpendicular to the surface of the panel for as long as possible. A smart solar tracker is a device that orients the solar panels in such a way that it maximizes the amount of incident solar radiation on the device. Most of the existing solar trackers have a computer or microcontroller that executes an algorithm with the bases of solar tracking [7].

Various studies on single and dual axis solar trackers have shown a gain of up to 30% for single axis solar trackers and up to 40% for dual axis solar trackers [8].

Two-axis solar trackers can maintain the optimum angle with respect to the sun to achieve maximum capture of solar radiation. Besides commercial devices, two-axis solar tracking devices have been developed on the Arduino platform due to its accessibility [9,10].

1.4 Evaporation of droplets on a surface

Undertaking research using the solar panel drying method involves understanding the mechanisms of drying water on a flat surface at any inclination angle.

The evaporation of drops is an essential physical process linked to many variables, separating its analysis between micro and macro drops (meaning "macro" drops those of the order of centimeters), facilitates the understanding of its evaporation, its dynamics, and the way in that it is linked to its role in the soiling effect. Although fouling does not depend only on the rainfall present in a region, it is a parameter that can improve the performance of the solar panel.

In general, a droplet is deformed by gravity when its size exceeds a characteristic size called the capillary length, approximately 2.7 mm; droplets that exceed this radius have a flatter shape than smaller droplets. On an inclined surface the shape of the droplets will be further deformed due to the competition between gravity and surface tension. In general, asphericity will increase with droplet size.

The evaporation time of a drop will depend on the diffusion coefficient, the vapor concentration at the liquid surface, the vapor concentration at infinity, the tilt angle of the substrate, the ambient temperature, as well as the initial temperature of the substrate and the water drop, several models have been developed to predict the evaporation time based on some of the mentioned properties [11].

The water droplets that include particles will generate a "coffee stain" effect, where the impurities present in a drop precipitate, accumulating mainly in the contact edges of the drop with the substrate [12]. Figure 1 shows the generation of stains by drying the drops.

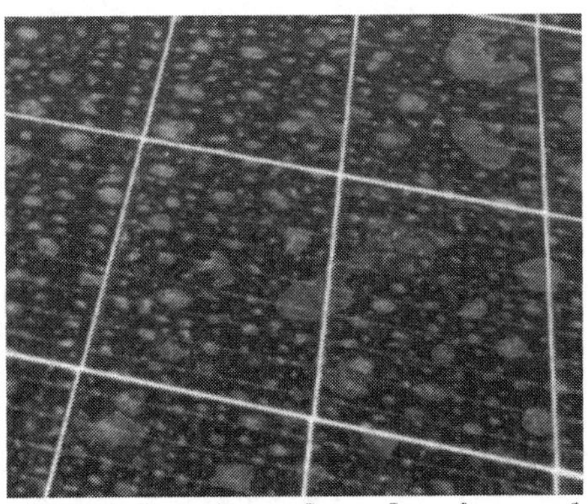

Figure 1. Formation of a "coffee stain" on the surface of a solar panel in natural rain conditions.

2. Experimental details

2.1 Experimentation device

To simulate the environmental conditions in which a solar tracker works, an experimental device was developed that allows us to control wind, rain and solar irradiance conditions.

The experimental device has a fan to simulate the wind in a drying process, this is fixed in front of a mini solar tracker by means of a clamp and can be removed while the rain process is carried out. At its maximum power, the wind speed reaches 1,525 m/s (5.49 km/h), taking measurements at 20 cm away from the fan. On the Beaufort scale for wind speed, a calm or light wind between 2 and 12 km/h is common for an average day and in this case, it is used for drying processes [13].

Lighting is provided by halogen lamps. In the experimental device, the halogen lamps are located 30 cm above the center of the surface of the solar tracker, when it is in its horizontal position. At this distance, the irradiance of the incident light generated by a single halogen lamp was measured, resulting in 252.3 W/m^2.

To simulate rain on the solar panel, a pop-up sprinkler fixed to the top of the structure was implemented.

The experimentation process was automated through the implementation of the Arduino development platform, an Arduino Mega 2560 was used to control all the functions of the device. The device's location is closed with transparent acrylic screen to minimize the influence from outside its cage. Figure 2 shows the dimensions and scheme of the developed device.

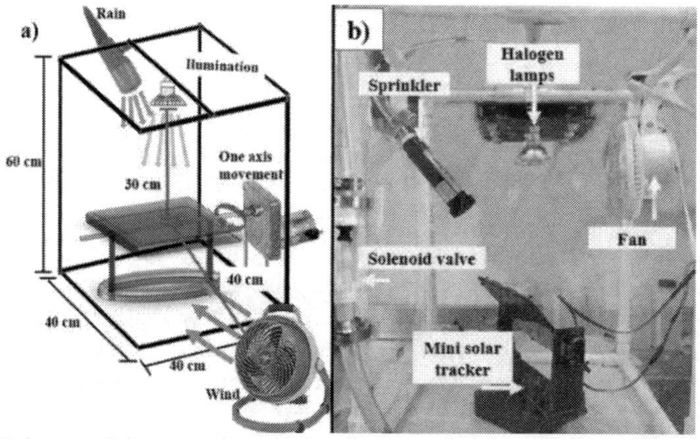

Figure 2. a) Scheme of the experimentation device and its dimensions, b) photo of the experimentation device.

2.2 Water drying process on solar panel surfaces

The solar cells are covered with an encapsulant that prevents the interaction of the cell with the environment, in turn, the solar cell encapsulant is protected by a layer of glass that is the one that interacts directly with the environment. Different types of glass are used as a protection layer, including common glass (also known as silica), tempered glass and monolithic glass [14]. As the glass layer interacts with rain, dust and wind, it is possible to use ordinary glass to simulate the drying process suffered by the surfaces of the solar panels. Cleaning and drying process is simulated to establish a drying protocol that allows the panel's sunscreens optimal inclination position to reduce the stains caused by water droplets when drying.

Although there are many variables that can affect the cleaning of solar panels, it is necessary to limit the experiments to various stages in which the conditions are controlled, for this reason, for the first stage of experimentation, only tap water was used, from the flow of drinking water, to simulate drying processes at various inclinations of the solar tracker.

The experimentation device allows to carry out the glass fouling process, which, as already mentioned, is carried out only with water, wetting it at an angle of inclination defined between 0° and 90°, defining the horizontal position of the follower as 0° and the vertical position as 90°.

Once the angle at which the experiment will be carried out has been selected, to generate the watermarks, a "rain" process is carried out, where the glass is sprayed for 10 seconds. At the end of the spraying time, lighting begins with or without wind, for a period of 60 minutes. As time passes, we can see that the water droplets gradually evaporate, leaving traces of the salts and minerals that the tap water contains [15]. By repeating the procedure and varying the angle of inclination, we can study the formation and density of these stains in search of locating the optimum drying position where the density of stains per area is lower.

The drying protocol (in passive cleaning) was varied in four experiments, the room temperature was kept constant at 24 °C, the active sprinkler time was 10 seconds and 2 mm thick glass was used, cut in a square shape with dimensions of 15 x 15 cm. Prior to the experiments the surfaces were carefully cleaned with acetic acid diluted in water and dried with microfiber.

Four different drying processes were carried out, with the parameters shown in table 1. In process 1 the fan was not activated, and it was left to dry only with lighting at an irradiation of 252.3 W/m^2, in process 2 the light intensity remained unchanged, but the fan was activated, with a wind speed of 1,525 m/s. For process 3, the irradiation was doubled to 504.6 W/m^2 and the fan was kept off, while in process 4 the light intensity was maintained at 504.6 W/m^2 and the wind speed at 1,525 m/s. In each of these drying processes, 17 angle variations were produced, starting from the horizontal position at 0°, increasing 5° per experiment, until completing the position close to the vertical at 80°.

After the drying process, a 6 x 5 cm area was selected on the glass, where photographs of this area were captured to carry out the fouling analysis. Finally, the images were analyzed using the ImageJ software [16], by a particle analysis process, in which the area occupied by the stains on the glass was measured.

Table 1. Varied parameters in the drying processes.

Protocol	Irradiation (W/m^2)	Wind velocity (m/s)
1	252.3	0
2	252.3	1.525
3	504.6	0
4	504.6	1.525

For a second series of experiments, a cleaning process under rain is proposed where the panel has an active role (active cleaning). It is proposed to generate a 30° agitation process with a frequency of 2 seconds, while it is in a rainy situation the solar tracker will activate a shaking mechanism (rocking movement) when rain is detected. In this series of experiments, angles close to the horizontal were prioritized for the drying phase, varying from 0° to 35°, with a variation of 5° between experiments, considering that the horizontal position of the panel is the least favorable for generating cleaning under rain and wind. It is convenient to measure the effect that rocking movement will have at angles close to the horizontal. The wind and irradiance parameters were 2.87 m/s and 504.6 W/m^2 respectively. For this series of experiments glass substrates were wet with a thin layer of water, then each substrate was filled with 1.4g of sprinkled sand, generating a layer of cemented sand after the thin layer of water is evaporated in a gas extraction hood at a room temperature of 24°, always taking care to avoid contamination of the substrates. With the described substrate preparation process the experiments simulate a more realistic situation compared with the accumulated soiling in a real environment.

Simulation of rain and the drying process was experimented with and without oscillations to compare results of a passive and active cleaning. In the rain process the active sprinkler time was 5 seconds, drying protocol and the photograph analysis were almost the same than the first experiments series.

3. Results and discussion

The stains patterns obtained in the first series of experiments (passive cleaning) were analyzed in ImageJ by image processing, calculating the average size of the particles (stains) that water forms on the substrates when it dries. In the image analysis, two kinds of areas are distinguished, those that are noticeably stained (shaded area) and those that are not shaded, as shown in figure 3. The shaded area is the result of the previously described process of coffee stains formation, were the particle impurities in the water agglomerate in the line of evaporation of water droplets. The results, as a percentage of shaded area, for each of the processes of drying in passive cleaning experiments are shown in figure 4. We can highlight that in the processes where wind was not used, the total shaded area of the substrates is slightly less, which may be due to a wind-generated drag effect on the water droplets, generating larger spots, an effect that is noticeable mainly at angles close to the horizontal position.

In figure 4, analyzed images were captured under a standardized method which enables high repeatability, focusing a selected area in each substrate. Each experiment was repeated four times, so each value obtained as a result is the average of four analysis. The exponential fitting curves have a standard deviation of 4.2% for protocol 1, 6.28% for protocol 2, 4.65% for protocol 3 and 6.58% for protocol 4.

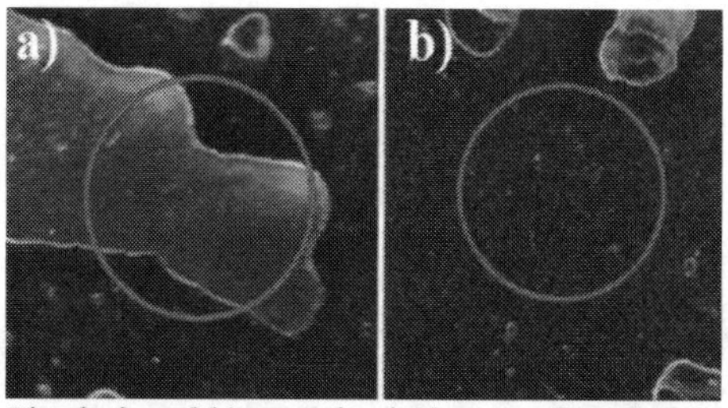

Figure 3. a) Shaded and b) not shaded area after the drying process.

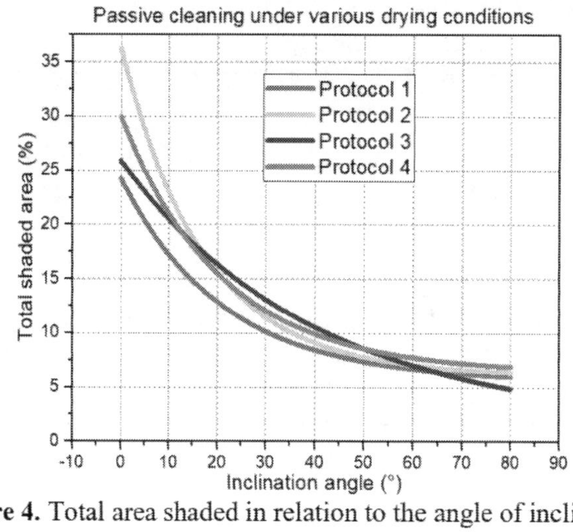

Figure 4. Total area shaded in relation to the angle of inclination.

The total shaded area analysis for the agitation process (rocking movement during active cleaning) is shown in figure 5, these experiments were proposed for angles close to the horizontal position, in this case it is notorious that the drying processes without oscillation (passive cleaning) show more dirt, which means a favorable result for the active cleaning process. The active cleaning process increases the contact time of the water layer with the surface of the substrate, allowing dirt to be removed more effectively. Standard deviation for passive cleaning exponential fitting is 5.72% and 4.2% for active cleaning.

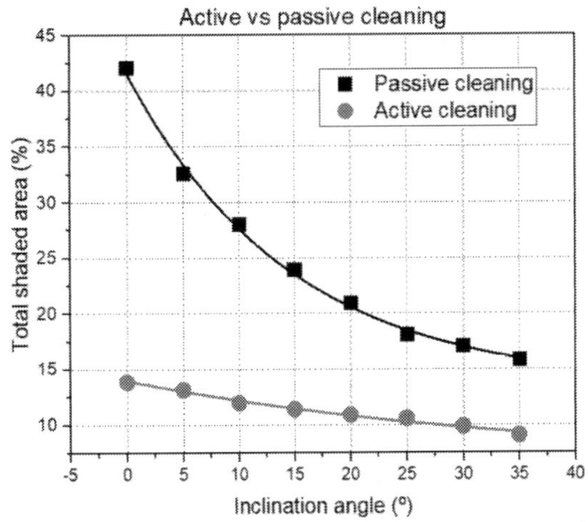

Figure 5. Comparison between passive and active washing.

A comparison of the IV (current-voltage) curves of a 5V commercial mini solar cell was made, when it has a dirt grade similar to that of the experiments prepared with sand. As a result, a reduction in the output current close to 50% was obtained, like the energy losses that occur in desert areas due to soiling.

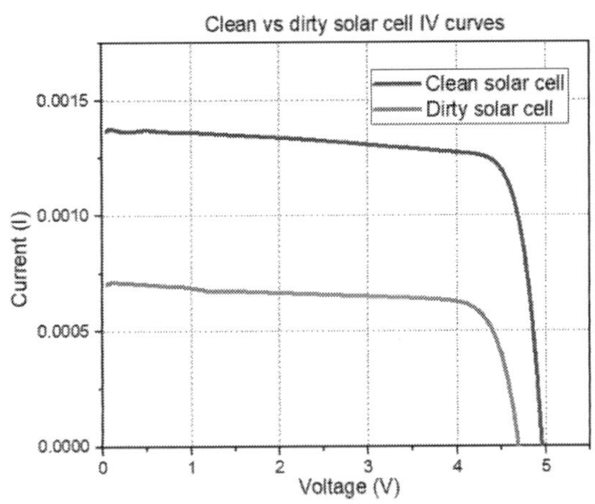

Figure 6. IV curves of a clean and dirty solar cell.

4. Conclusions

In this work, active and passive washing methodologies for solar panel surfaces were experimented, taking advantage of the positioning capabilities of a solar tracker. Previous articles studied the relationship between soiling and its behavior in rainy conditions, but the capability to use solar trackers as a support for cleaning has not been studied.

A notable result of the first series of experiments is that the highest slope of the graphs is between 0° and 30°, which indicates that it would not be enough to take the panels to an angle of inclination close to 30° to obtain an acceptable drying with a significantly reduced number of spots. Finally, the experiments with active cleaning show a notable improvement in the cleaning process under rain, this result that can be implemented to take advantage of weather conditions with moderate or light rainfall. Active and passive cleaning are novel cleaning methods with the capacity to be implemented in weathers with enough rainfall periods.

Acknowledgments

This work received support from the CONACYT-SENER-Energy Sustainability Sector Fund through Project 207450, "Centro Mexicano de Innovación en Energía Solar (CeMIE-Sol)" and particularly within Strategic Project No. 91 "Development of a smart solar tracker prototype with communication functions and strategic control".

References

[1] Schill C., Brachmann S., Koehl M., "Impact of soiling on IV-curves and efficiency of PV-modules", Solar Energy, vol 112, pp. 259-262, 2015.

[2] Mallikarjun G., Gita J., Umayal R., Ashwini R., "A Comparative Study of Dust Cleaning Methods for the Solar PV Panels", Advanced Journal of Graduate Research, vol 1, pp. 24-29, 2017.

[3] Hajighorbani S., Radzi M. ,Ab KadirM. ,Shafie S., Khanaki R. ,Maghami M., "Eva-luation of Fuzzy Logic Subsets Effects on Maximum Power Point Tracking for Photovoltaic System". International Journal of Photoenergy, 2014.

[4] Kimber, A., Mitchell, L., Nogradi S., Wenger H., "The effect of soiling on large grid-connected photovoltaic systems in California and the Southwest Region of the United States", Conference Record of the 2006 IEEE 4th World Conference on Photovoltaic Energy Conversion, vol 2, pp. 2391-2395, 2006.

[5] Asl-Soleimani E., Farhangi S., Zabihi M., "The effect of tilt angle,air pollution on performance of photovoltaic systems in Tehran", RenewEnergy, vol 24, pp. 459–468, 2001.

[6] M. Mazumder, J. Yellowhair, J. Stark, C. Heiling, J.Hudelson, F. Hao, H. Gibson, M. Horenstein, "Optical and Adhesive Properties of Dust Deposits on Solar Mirrors and their Effects on Specular Reflectivity and Electrodynamic Cleaning for Mitigating Energy-yield Loss", High and Low Concentrator Systems for Solar Energy Applications IX, vol. 9175, 2014.

[7] Rezoug M. R., Chenni, R., "The Optimal Angles of a Dual-axis Tracking System by Pre-programmed Method Using a Microcontroller". International Journal of Engineering Sciences & Research Technology, vol 7(3), pp. 710-726, 2018.

[8] Dhanabal.R., Bharathi.V., Ranjitha.R., Ponni.A., Deepthi.S., and Mageshkannan.P. 2013. "Comparison of efficiencies of solar tracker systems with static panel single axis tracking system and dual axis tracking system with fixed mount." International Journal of Engineering and Technology (IJET), vol 5, pp. 1925-1933, 2013.

[9] Morón Carlos, Ferrández Daniel, Saiz Pablo, Vega Gabriela and Díaz Jorge Pablo, "New prototype of photovoltaic solar tracker based on arduino", Energies, vol. 9, pp. 1-13, 2017.

[10] Fadhil Muthna Jasim, Fayadh Rashid Ali and Wali Mousa K, "Design and implementation of smart electronic solar tracker based on Arduino", Telkomnika (Telecommunication Computing Electronics and Control), vol. 17, pp. 2486-2496, 2019.

[11] Picknett R., Bexon R., "The evaporation of sessile or pendant drops in still air", J. Colloid Interface Sci, vol 61, pp. 336–350, 1977.

[12] H. B. Eral, D. J. C. M., J. M. Oh, "Contact angle hysteresis: A review of fundamentals and applications", Colloid and Polymer Science, vol 291, pp. 247-260, 2013.

[13] A.M. Da Silva, C.C. Young and S. Levitus, "Toward a Revised Beaufort Equivalent Scale", The American Journal of Sports Medicine, pp. 270-286, 1995.

[14] Mohammad Humood, Ali Beheshti Andreas A. Polycarpou,"Surface reliability of annealed and tempered solar protective glasses: Indentation and scratch behavior", Solar Energy, vol. 142, pp. 13-25, 2017.

[15] Mora Abrahan, Mahlknecht Jürgen, Rosales-Lagarde Laura and Hernández-Antonio Arturo, "Assessment of major ions and trace elements in groundwater supplied to the Monterrey metropolitan area, Nuevo León, Mexico", Environmental Monitoring and Assessment, vol. 189, 2017.

[16] C.A. Schneider, W.S. Rasband and K.W. Eliceiri, "NIH Image to ImageJ: 25 years of image analysis", Nature Methods, vol. 9, pp. 671-675, 2012.

AUTHOR INDEX